M000238345

Coin Tossing: The Hydrogen Atom of Probability
by Stefan Hollos and J. Richard Hollos

Abrazol Publishing
an imprint of Exstrom Laboratories LLC
662 Nelson Park Drive, Longmont, CO 80503-7674 U.S.A.

Publisher's Cataloging in Publication Data
Hollos, Stefan
Coin Tossing: The Hydrogen Atom of Probability / by Stefan Hollos and J. Richard Hollos
p. cm.
Includes bibliographical references.
Paper ISBN: 978-1-887187-38-1
Ebook ISBN: 978-1-887187-39-8
Library of Congress Control Number: 2019904573
1. Probabilities 2. Stochastic processes
I. Title. II. Hollos, Stefan.
QA274.H65 2019
519.2 HOL

2010 Mathematics Subject Classification. Primary 60-01, 62-01

About the Cover: Coin images from the United States Mint.

\

WORLD PEACE MUST DEVELOP
FROM INNER PEACE.
PEACE IS NOT THE ABSENCE OF VIOLENCE.
PEACE IS THE MANIFESTATION
OF HUMAN COMPASSION.

14th Dalai Lama of Tibet (Tenzin Gyatso) Coin.
Image credit: United States Mint.

Contents

1 Preface

This book is an expansion of the first edition of our Coin Toss book (**The Coin Toss: Probabilities and Patterns**), which was published only as an ebook, and released in September 2012, almost 7 years ago. Actually, the initial title was **The Coin Toss: The Hydrogen Atom of Probability**, which we soon thereafter changed because we thought the allusion to physics would be too esoteric for most readers. But we have since come around to the idea that the initial title was the better choice.

The big difference between this and the first edition is chapter 8 on run distributions. The chapter now has an extensive discussion of statistics on the number of runs in a binary sequence generated by fair coins, biased coins and binary Markov processes. It also looks at statistics for the number of runs of a given length. We also show how to find probability generating functions and expectations for the length of the longest run in a binary sequence. Another feature of this new edition is a chapter containing 26 problems and solutions.

The book contains material for both the beginning student and the advanced researcher. We suspect that the beginner will find some of the material quite difficult and not accessible on a first reading. This is a book that needs to be read more than once. There is more material here than anyone could absorb on a first reading. We hope that researchers on the other hand find the book to be a valuable reference and a stimulus for new research.

American Buffalo One Ounce Gold Proof Coin.
Image credit: United States Mint.

Most physicists would probably agree that quantum mechanics could never have been discovered without the hydrogen atom. The hydrogen atom is the simplest of all the elements with just one electron surrounding a nucleus composed of just one proton (there are two other forms of hydrogen that have neutrons in the nucleus). This simple arrangement gives hydrogen a clean emissions spectrum which allowed physicists to find formulas for the emission lines. Eventually Niels Bohr (1885-1962) came up with a model explaining the emission lines and the formulas. The model was developed further by other physicists and eventually evolved into what we now call quantum mechanics. Being able to start with the simple structure of the hydrogen atom is what made this all possible.

The coin toss is to probability theory what the hydrogen atom is to quantum mechanics. It is the simplest random event that you can imagine. There are only two possible outcomes: heads or tails. This simplicity means that many questions about coin tossing can be asked and answered in great depth. The simplicity also opens the road to more advanced probability theories dealing with events with an infinite number of possible outcomes. Without the example of simple random events like tossing coins or rolling dice it is very likely that probability theory would have taken much longer to develop. Indeed the beginnings of mathematical probability theory are usually traced to the analysis of a dice game by Blaise Pascal (1623-1662) and Pierre de Fermat (1607-1665) in the mid-seventeenth century. The roll of a die is very similar to the toss of a coin except that there are 6 possible outcomes instead of only 2. Most of the things that we will talk about concerning coins can be easily extended to dice.

Keep in mind as you read what follows, that the coin toss is really just a metaphor for a random event that has only two possible outcomes. The actual tossing of a real coin is just one way to realize such an event. There are many examples of things that are equivalent to a coin toss:

- Will the stock market close up or down tomorrow?

- Will a dice roll come up with an even or odd number?

- Will we make contact with extraterrestrials within the next ten years?

- Will the Denver Broncos win their next game?

- Will a car drive by in the next minute?

- Will the Republican or Democratic candidate win the next election?

- Will tomorrow be sunny or cloudy?

- Will tomorrow be warmer than today?

- Will my medical test result be negative or positive?

- Will I enjoy this movie?

- Will the next joke be funny?

- Will the Earth's average temperature go up next year?

You can probably think of hundreds of other examples.

In the rest of this introduction we will take a detailed look at what happens when you toss a coin multiple times. This will lead in the next chapter to the subject of probability distributions that are associated with tossing a coin. We will show how many of the questions that can be asked about coin tossing are answered by specific distributions. The following chapter discusses some strategies for betting on coin toss games. The next chapter introduces the random walk interpretation of a series of coin tosses, and how the gambler's ruin problem can be viewed as a random walk with absorbing boundaries. This is followed by a chapter on runs and patterns in a sequence of coin tosses. The statistics of runs and patterns is useful for playing games where at least some aspect of the game can be modeled as a coin toss. Some forms of investing and trading fall in this category. Runs and patterns also come up in areas like reliability analysis and testing for defects in a manufacturing process. The next chapter shows how runs and patterns can be analyzed as Markov chains which is a very powerful technique that can be used to look at patterns of arbitrary complexity. We end with a chapter on the probability distribution for the longest run, followed by an appendix that reviews discrete probability.

So let's start by asking what happens when you toss a coin multiple times. If you toss a coin n times how many different strings of heads and tails can you get? For one toss you get H or T. Two tosses gives you TT, TH,

HT, or HH. Every additional toss doubles the number of possible strings. So for n tosses there are 2^n possible strings. This gets large very quickly. For 10 tosses you have $2^{10} = 1024$ possibilities. Out of all these possible strings what is the probability of getting a particular one?

To answer that you need to know the probability of getting heads or tails on a single toss. Let's assume the probability of heads is p and tails is $q = 1 - p$ for every toss. You can also assume that what happens on one toss cannot affect any subsequent toss. This makes the tosses independent random events and simplifies the probability calculation. The probability of a particular result for a series of independent random events is just the product of the individual probabilities. Take for example the following sequence of tosses:

```
H H H T T H H H H T T T H T H
p p p q q p p p p q q q p q p

H H H H T H T T H T T H H T T
p p p p q p q q p q q p p q q
```

The probability of each toss is shown on the second line. Since the tosses are independent, the probability of the sequence is just the product of the probabilities for each toss. There are a total of 30 tosses with 17 heads and 13 tails so the probability is $p^{17}q^{13}$. In general if you toss the coin n times, getting k heads and $n - k$ tails, then the probability will be

$$p^k q^{n-k} = p^k(1 - p)^{n-k} \qquad (2.1)$$

If the coin is fair then $p = q = 1/2$. In this case the probability of the above sequence or any sequence of 30 tosses is $1/2^{30}$. When the probability of heads and tails is the same, the probability of any sequence of n tosses will be $1/2^n$. Every sequence has the same probability. Now you may be wondering, how is this possible? Isn't it more likely to get about the same number of heads and tails with a fair coin than it is to get say all heads? But we just said that all sequences have the same probability. What gives?

The problem is that while there is only one sequence that has all heads, there are many sequences that have the same number of heads and tails. What we really need to compare is the probability of getting the single sequence of all heads and the probability of getting any one of the many sequences with equal numbers of heads and tails.

How many sequences of $2n$ tosses (we use $2n$ instead of n since the number of tosses has to be even to get the same number of heads and tails and $2n$ is explicitly even) have an equal number of heads and tails? This is equivalent to asking for the number of ways you can divide the set of $2n$ tosses into two equal subsets of size n. One subset has the heads and the other the tails. In general if you have $2n$ objects and you want to pick out n of them, there are $2n$ ways to pick the first, $2n - 1$ ways to pick the second, and so on down to $n + 1$ ways to pick the last. The total number of ways is then [1]

$$2n(2n-1)(2n-2)\cdots(n+1) = \frac{(2n)!}{n!} \qquad (2.2)$$

But not all of these ways will end up with a unique set of objects since the same objects can simply be picked in a different order. A given set of n objects can be picked in $n! = n(n-1)(n-2)\cdots 1$ ways so if $C(2n,n)$ is the number of unique ways of picking n objects out of $2n$ then $C(2n,n)n!$ must equal equation 2.2 and we have:

$$C(2n,n) = \frac{(2n)!}{n!n!} \qquad (2.3)$$
$$= \binom{2n}{n}$$

Using this result you can then see that the probability of getting the same number of heads and tails with a fair coin is $C(2n,n)/2^{2n}$ while the probability of getting all heads is $1/2^{2n}$. For large n, $C(2n,n)$ will be very much larger than 1 so there is a much higher probability of an equal number of heads and tails than of all heads.

The fact is that comparing the probabilities of two particular sequences is oftentimes not very meaningful. A particular sequence with k heads and $n-k$ tails will have the same probability as any other sequence with the same number of heads and tails. Using arguments similar to

---

[1]The symbol '!' is called 'factorial', and for an integer n, is a product of the integers $n(n-1)(n-2)\cdots 1$.

above you can show that the number of ways of getting k heads and $n-k$ tails in a sequence of n tosses is given by [2]

$$\binom{n}{k} = \frac{n!}{k!(n-k)!} \qquad (2.4)$$

The probability of getting a particular one of these sequences is given by equation 2.1 so the probability of getting any one of the sequences is:

$$P(n,k) = \binom{n}{k} p^k (1-p)^{n-k} \qquad (2.5)$$

This is the probability of getting k heads in a series of n coin tosses where the order in which they appear makes no difference. It is a probability distribution called the binomial distribution and it is just one of the many probability distributions associated with coin tossing.

---

[2]Equation 2.4 is also known as a binomial coefficient.

Sacagawea Golden Dollar Coin.
Image credit: United States Mint.

3 Probability Distributions

Some of the probability distributions related to coin tossing come from answering questions like: what is the probability that a coin has to be tossed n times before the first head appears. Others relate to questions about what happens when the number of tosses goes to infinity. Most of these distributions will be discussed below, starting with the simple discrete distributions and moving on to continuous probability density functions.

Before we begin let's clear up some terminology. When we talk about a probability distribution what we mean is a way of assigning a probability to the outcome of a random event. If there are only a finite number of possible outcomes, such as in a coin toss or a dice roll, then the probability distribution is called a probability mass function (pmf). This is also true if the number of outcomes is infinite but each outcome can be assigned an integer value. An example of this is the number of tails that appear before the first head in a sequence of coin tosses. If the outcomes form a continuous range of values, such as a person's weight, then the probability distribution is called a probability density function (pdf). In this case the probability of any one specific value is zero and you can only assign a probability to a range of values. You do this by integrating over the range. In many cases we will refer to both a pmf and a pdf as a probability distribution or just a distribution.

3.1 Bernoulli Distribution

The simplest distribution associated with coin tossing is the Bernoulli distribution, named after Jacob Bernoulli (1654-1705).

Figure 3.1: Jacob Bernoulli (1654-1705). Image credit: Wikimedia.org

It is the distribution for a single toss of the coin which we will represent by the random variable T. A random variable assigns a number to each outcome of a random event. For a single toss the only outcomes are heads or tails. For heads we will let $T = 1$ and for tails $T = 0$.

Since there are only two values the probability distribution can be stated by just listing them as follows:

$$
\begin{aligned}
P(T = 1) &= p \\
P(T = 0) &= 1 - p
\end{aligned}
\qquad (3.1)
$$

Or it can be written more compactly as:

$$
P(T) = p^T (1 - p)^{1-T} \qquad (3.2)
$$

The expectation (mean) and variance of T are: [1]

$$
\begin{aligned}
E[T] &= p \\
\mathrm{Var}[T] &= p(1 - p)
\end{aligned}
\qquad (3.3)
$$

These are of little use by themselves but they are useful for characterizing multiple tosses which leads us next to the binomial distribution.

3.2 Binomial Distribution

If you toss a coin n times and you want to know the probability of getting k heads in any order then the Binomial

[1] For a definition of mean and variance see the Review of Discrete Probability in Appendix A.

distribution is your answer. Let S_n be the random variable representing the number of heads that appear when a coin is tossed n times. If T_i is the Bernoulli variable for the ith toss then S_n is equal to the sum of all the T_i.

$$S_n = \sum_{i=1}^{n} T_i \qquad (3.4)$$

S_n is called a binomial random variable. Its value can range from 0 to n. The probability that S_n equals k is equal to the probability of getting any sequence of n tosses with k 1's. This probability was worked out in the introduction and is given by equation 2.5 so that

$$P(S_n = k) = b(n, k) = \binom{n}{k} p^k (1 - p)^{n-k} \qquad (3.5)$$

This is called the binomial distribution. To get the probability that S_n has a value less than or equal to m you just sum this equation from 0 to m

$$P(S_n \leq m) = \sum_{k=0}^{m} \binom{n}{k} p^k (1 - p)^{n-k} \qquad (3.6)$$

When n is large this sum can be hard to evaluate. Luckily the binomial distribution can be well approximated by a normal distribution. We will show how to do this later in section 3.4 on the normal distribution. The QuantWolf Binomial Distribution Calculator allows you to evaluate equation 3.6 directly, and also provides the normal distribution approximation.

The probability generating function [2] for the binomial distribution is

$$G_n(z) = (pz + q)^n \qquad (3.7)$$

If you expand out the right side of this equation then the coefficient of z^k will be the probability that $S_n = k$ given by equation 3.5.

Probability generating functions are very useful for calculating the expectation, variance, and other moments of a random variable. The expectation, for example, is the first derivative of $G_n(z)$ with respect to z evaluated at $z = 1$. In this case, however, it is easier to use the fact that S_n is the sum of the Bernoulli random variables for each toss as shown in equation 3.4. Since the tosses are independent you can get the expectation and variance by just summing the expectation and variance for each toss. This gives:

$$\begin{aligned} E[S_n] &= np \\ \mathrm{Var}[S_n] &= np(1-p) \end{aligned} \qquad (3.8)$$

These equations tell you that when $p = 1/2$ you can expect about half the tosses to be heads (1) and the variance will be a maximum of $n/4$. This means that most of the

---

[2]For a review of generating functions, see Herbert Wilf's book *generatingfunctionology* which you can download from his website at: https://www.math.upenn.edu/ wilf/

time in a series of n tosses with $p = 1/2$ the number of
1's will fall between $n/4$ and $3n/4$.

Figure 3.2 shows plots of the binomial distribution for $n = 10$ with $p = 1/2$ on the top and $p = 4/5$ on the bottom.
Note that with $p = 1/2$ the distribution is symmetric.
For a fair coin, the probability of getting 2 heads in 10
tosses is the same as the probability of getting 8 heads.
This will be true for any value of n so that in general
$P(S_n = k) = P(S_n = n - k)$. When n is even the
distribution will peak at $S_n = n/2$ and when n is odd there
will be two peaks at $S_n = (n-1)/2$ and $S_n = (n+1)/2$.

With $p = 4/5$ on the other hand the distribution is skewed.
The distribution will be skewed for any value of n when-
ever $p \neq 1/2$. The probability that $S_n = k$ will equal the
probability that $S_n = n - k$ if p is replaced by $1 - p$. For
example the probability of getting 2 heads in 10 tosses
when $p = 1/5$ is the same as the probability of getting
8 heads in 10 tosses when $p = 4/5$. For a skewed distribution the peak will always be between $(n + 1)p - 1$ and
$(n + 1)p$. If $(n + 1)p$ is an integer then there will be a
peak at both $(n + 1)p - 1$ and $(n + 1)p$.

3.3   Beta Distribution

To calculate any probability associated with a coin toss
you need to know the probability of getting heads on a
single toss which we have been calling p. So far we have
taken the value of p as a given. But what if it's not?
What do you do then? If we are talking about tossing a

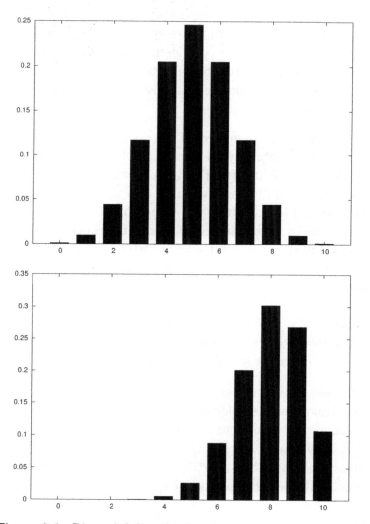

Figure 3.2: Binomial distribution for $n = 10$ with $p = 1/2$ on the top and $p = 4/5$ on the bottom.

real coin that is balanced and tossed in a way that does not favor one side or the other then we can assume $p = \frac{1}{2}$. If the possibility of a bias is suspected so that either heads appears more often, $p > \frac{1}{2}$, or tails more often, $p < \frac{1}{2}$, then the value of p has to be determined experimentally. The experiment is simple, you toss the coin n times and count how many heads appear.

For a particular sequence of n tosses in which k heads appear, the probability is $P(p) = p^k (1 - p)^{n-k}$. This is a function of the unknown heads probability p. The obvious thing to do is find the value of p that maximizes the function. This can be done using the standard calculus technique of taking the derivative with respect to p, setting it equal to zero and solving for p. If you do this, you get what seems like the obvious answer of $p = k/n$. In other words p is just equal to the fraction of heads that occur in n tosses. This is called the maximum likelihood estimate for p. As n gets larger and larger, the estimate will become more accurate and in the limit of an infinite number of tosses it will become exact.

For any finite number of tosses there will still be some uncertainty in the value of p and it would be nice to have some idea of how large the uncertainty is. One way to do this is by using what is called Bayesian inference, named after Thomas Bayes (1701-1761), which will give us a probability distribution for p called the beta distribution.

In general Bayesian inference relates the probability of a hypothesis given some data to the probability of getting that data given that the hypothesis is true. If H represents the hypothesis and D the data, then Bayes' Theorem sim-

Figure 3.3: Thomas Bayes (1701-1761), purportedly. Image credit: Wikimedia.org

ply says that

$$P(H|D) = \frac{P(D|H)P(H)}{P(D)} \qquad (3.9)$$

In our case the hypothesis is that the probability of heads is equal to p which we denote as $H = p$. The data is just the sequence of heads and tails that result from flipping the coin n times. The term $P(D|H)$ is called the likelihood function. It is the probability of getting the data given the hypothesis which we determined above to be

$$P(D|H) = p^k(1 - p)^{n-k} \qquad (3.10)$$

The term $P(H)$ is called the prior probability of $H = p$.

It represents what is known about the value of p before the n tosses. It could for example come from an estimate of p obtained from a previous set of n tosses. In other words equation 3.9 can be applied recursively. The value of $P(H|D)$, called the posterior probability of $H = p$ given the data, can become the prior probability $P(H)$ for a subsequent round of another n tosses.

If there have been no previous tosses and nothing is known about the coin then $P(H)$ should be a uniform distribution, i.e. $P(H) = 1$ for $0 \leq p \leq 1$. This means all values of p are equally likely from $p = 0$ where we have a coin with no heads (both sides are tails) to $p = 1$ where both sides of the coin are heads. So for a uniform prior distribution, equation 3.9 becomes

$$P(H|D) = \frac{p^k(1-p)^{n-k}}{P(D)} \qquad (3.11)$$

The value of $P(D)$ comes from the normalization condition required to make $P(H|D)$ a proper probability distribution

$$\int_0^1 P(H|D)\,dp = 1 \qquad (3.12)$$

This condition implies that

$$P(D) = \int_0^1 p^k(1-p)^{n-k}\,dp \qquad (3.13)$$

The integral is a special function called a beta function

which is generally defined as

$$B(a,b) = \int_0^1 p^{a-1}(1-p)^{b-1}\, dp \qquad (3.14)$$

The beta function can be expressed in terms of another function called a gamma function as

$$B(a,b) = \frac{\Gamma(a)\Gamma(b)}{\Gamma(a+b)} \qquad (3.15)$$

where $\Gamma(x)$ is defined as $(x > 0)$

$$\Gamma(x) = \int_0^\infty t^{x-1}e^{-t}\, dt \qquad (3.16)$$

When x is an integer $\Gamma(x)$ becomes simply the factorial function $\Gamma(x) = (x-1)!$.

In our case the parameters of the beta function are $a = k+1$ and $b = n-k+1$ so we have

$$B(k+1, n-k+1) = \frac{\Gamma(k+1)\Gamma(n-k+1)}{\Gamma(n+2)} \qquad (3.17)$$

All the gamma function arguments are integers which makes this expression a ratio of factorials that can be simplified as

$$B(k+1, n-k+1) = \frac{1}{(n+1)\binom{n}{k}} \qquad (3.18)$$

This then becomes the value of $P(D)$ in equation 3.11 giving us finally

$$P(H|D) = (n+1)\binom{n}{k}p^k(1-p)^{n-k} \qquad (3.19)$$

This looks remarkably similar to the binomial distribution. There are some significant differences however. The binomial distribution is a function of n and k with the value of p as a given. It gives you the probability of getting k heads in n tosses given a particular value of p. $P(H|D)$ on the other hand is a function of p with the values of n and k as a given. It gives you the probability that the probability of heads is p given that there were k heads in n tosses. In other words it is a probability distribution for the probability of heads. This probability distribution has a name, it is called the beta distribution. Using a maximum likelihood approach gave us the single point estimate $p = k/n$. With the Bayesian approach we have a probability distribution for p which determines the most likely value as well as the uncertainties. A plot of $P(H|D)$ for $n = 10$ and $k = 6$ is shown in figure 3.4. Notice that the peak is at $p = 0.6$ which is equal to the maximum likelihood value, but now that we have a distribution we can find the probability that p is in the range $0.5 \leq p \leq 0.7$ by integrating $P(H|D)$ over the range.

There are a few more things to say about the beta distribution before we move on. In general the distribution is

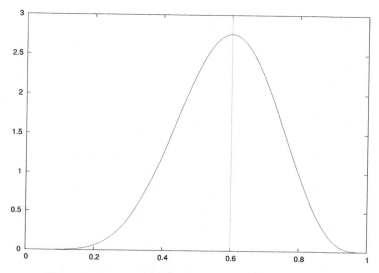

Figure 3.4: $P(H|D)$ for $n = 10$ and $k = 6$.

defined as:

$$\text{Beta}(p|a,b) = \frac{p^{a-1}(1-p)^{b-1}}{B(a,b)} \qquad (3.20)$$

where $B(a,b)$ is the beta function defined in equation 3.14. The mean and variance of the distribution are given by:

$$
\begin{aligned}
E[p] &= \frac{a}{a+b} \\
\text{Var}[p] &= \frac{ab}{(a+b)^2(a+b+1)}
\end{aligned}
\qquad (3.21)
$$

Another way to characterize the distribution is by its mode

which is given by

$$\text{Mode} = \frac{a - 1}{a + b - 2}$$

The mode is the value of p where the distribution has a maximum. The mean is equal to the mode only when $a = b$ in which case the distribution is symmetric. For the coin toss experiment $a = k + 1$ and $b = n - k + 1$ so the mode is k/n while the mean is $(k + 1)/(n + 2)$. Which of these is the best estimate to use for p? According to Laplace's *rule of succession* you should use the mean only if your prior state of knowledge is that both heads and tails have a nonzero possibility of occurring. In this case there is an implied set of two extra observations, one head and one tail, so the best point estimate is the mean, $p = (k + 1)/(n + 2)$. In the *complete ignorance* assumption there is no implied observation of a head and tail. For all we know the coin could have two heads or two tails before we see the result of any tosses. In the *complete ignorance* case the best estimate is the mode, $p = k/n$. This all gets into the technical details of how the Bayesian inference is carried out which we will not go into. In any case if n is large there should be little difference between using the mean or the mode to estimate p.

3.4 Normal Distribution

Here is a common question. If I toss a coin n times what is the probability that I get less than m heads? The answer is given by the binomial cumulative distribution function defined in equation 3.6. The problem is that the equation can be difficult to calculate when n is large (let's say greater than 100). The cause of the difficulty comes from evaluating the binomial coefficients in the equation. They become very large when n is large and the probabilities they multiply become very small. The problem has a long history. It was first tackled by Abraham de Moivre (1667-1754) in the second edition of his book (1738) *The Doctrine of Chances* (p. 235-243). His results were gen-

Figure 3.5: Abraham de Moivre (1667-1754). Image credit: University of York: Portraits of Statisticians, Wikimedia.org

eralized by Pierre-Simon, marquis de Laplace (1749-1827)

and the result is today known as the de Moivre–Laplace Theorem. The theorem says that for large n the bino-

Figure 3.6: Pierre-Simon, marquis de Laplace (1749-1827). Image credit: Jean-Baptiste Paulin Gurin, Wiki-media.org

mial distribution can be approximated as follows (with $q = 1 - p$)

$$\binom{n}{k} p^k q^{n-k} \approx \frac{e^{-\frac{(k-np)^2}{2npq}}}{\sqrt{2\pi npq}} \tag{3.22}$$

The expression on the right is the probability density function for the normal distribution. This probability density function is usually written in the following form

$$N(x|\mu, \sigma^2) = \frac{e^{-\frac{(x-\mu)^2}{2\sigma^2}}}{\sqrt{2\pi\sigma^2}} \tag{3.23}$$

The parameters μ and σ^2 are the mean and variance of the normal distribution.

$$\begin{aligned} E[x] &= \mu \\ \text{Var}[x] &= \sigma^2 \end{aligned} \tag{3.24}$$

By comparing equations 3.22 and 3.23 you can see that the binomial distribution probabilities can be approximated by a normal probability density function with mean and variance equal to the mean and variance of the binomial distribution. The approximation works best when p is close to $1/2$ and n is larger than 100.

Deriving the approximation in equation 3.22 is somewhat long and tedious. One way is to start with the approximation of the factorial function due to James Stirling (1692-1770). Stirling's approximation says that

$$n! \approx \sqrt{2\pi n} \left(\frac{n}{e}\right)^n \tag{3.25}$$

where $e = 2.718281828459\ldots$ is the base of the natural logarithm. Using this for the factorials in equation 3.22 and applying a long series of simplifications and limit arguments will eventually give you the approximation. We

will leave out the details and just use the approximation
as is.

Now to calculate the probability of getting less than or
equal to m heads in n tosses you can change the sum
in the binomial cumulative distribution function, equation
3.6 into an integral over the normal probability density
function. Figure 3.7 shows an example for $n = 20$ and
$p = 1/2$. Both the binomial distribution and its normal
approximation are shown. The probability of getting less
than 6 heads is equal to the area under the first six boxes
in the plot and this can be approximated by the corresponding area under the plot of the normal distribution.
The area under the normal distribution is found by integrating the normal probability density function from $-\infty$
to 6.5 or in the general case from $-\infty$ to $m + 1/2$.

To do the calculations it is usually best to work with a
standard normal distribution which has a probability density function given by:

$$N(z|0,1) = \frac{e^{\frac{-z^2}{2}}}{\sqrt{2\pi}} \qquad (3.26)$$

The standard normal is a normal distribution with a mean
of 0 and a variance of 1. You convert an integral of
$N(x|\mu, \sigma^2)$ into an integral of $N(z|0,1)$ by transforming
the x limits to z limits as follows:

$$z = \frac{x - \mu}{\sigma} \qquad (3.27)$$

Normal distribution integrals can generally not be evaluated in closed form but tables exist for the standard normal where you can look up the value and there are numerous ways to calculate it numerically, such as is done on the QuantWolf Normal Distribution Calculator.

As an example let's look at the case where $n = 50$, $m = 20$, and $p = 1/2$. We want the probability that in 50 tosses of a fair coin, there will be no more than 20 heads. An exact calculation of the probability, using equation 3.6, gives $P(S_{50} \leq 20) = 0.101319$. For the normal approximation we use $\mu = np = 25$, $\sigma^2 = npq = 50/4$ and integrate from $-\infty$ to $20 + 1/2$. Converting to a standard normal, the integration will be from $-\infty$ to

$$\frac{20.5 - 25}{\sqrt{12.5}} = -0.9\sqrt{2} \qquad (3.28)$$

This gives a probability of 0.101546 which is very close to the exact calculation.

3.5 Geometric Distribution

Suppose you toss a coin five times and a head doesn't show up until the fifth toss. What is the probability of this happening? In general what is the probability of tossing a coin m times with no head appearing until the last toss. The answer to this question is given by the geometric distribution.

The geometric random variable X is equal to the number

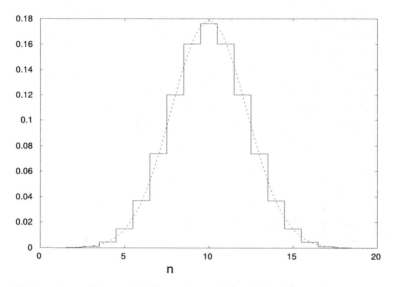

Figure 3.7: Binomial distribution (solid line) with $n = 20$, $p = 0.5$, and its normal distribution approximation with $\mu = np = 10$, and $\sigma^2 = npq = 5$.

of tosses that occur until the first head appears. For example with the sequence TTTTH X would have the value 5. In general for X to have the value m we must have $m - 1$ tails followed by a head so the probability must be:

$$P(X = m) = p(1 - p)^{m-1} \qquad (3.29)$$

This is known as the geometric distribution. To find the probability that X has a value less than or equal to m you just sum this from 1 to m. The sum can be expressed nicely as:

$$P(X \leq m) = \sum_{k=1}^{m} p(1 - p)^{k-1} = 1 - (1 - p)^{m} \quad (3.30)$$

If you want the probability that X has a value greater than m then just subtract the above sum from 1, $P(X > m) = 1 - P(X \leq m)$.

The probability generating function for the geometric distribution is

$$G(z) = \frac{pz}{1 - qz} \qquad (3.31)$$

from which you can readily get the probabilities in equation 3.29. The expectation and variance of X are also easily found from the generating function [3], they are:

$$E[X] \;=\; \frac{1}{p} \qquad\qquad (3.32)$$

$$\mathrm{Var}[X] \;=\; \frac{1-p}{p^2}$$

For small values of p the expectation and variance are large since you will, on average, have to toss many times before getting a 1 but there is always the chance that a 1 will appear quickly.

3.6   Negative Binomial Distribution

The geometric random variable can be generalized in a couple of different ways. We can for example ask for the probability that it takes m tosses to get r 1's. If X is the random variable and $r = 4$ then for the sequence 110001001 X would have the value 9. In general the probability that X has the value m is given by

$$P(X = m) = \binom{m-1}{r-1} p^r (1-p)^{m-r} \qquad (3.33)$$

where m must be greater than or equal to r. This is one version of what is known as the negative binomial distribution. To find the probability that X has a value

---

[3]See the Review of Discrete Probability in Appendix A.

less than or equal to m you sum this from r to m

$$P(X \leq m) = \sum_{k=r}^{m} \binom{k-1}{r-1} p^r (1-p)^{k-r} \qquad (3.34)$$

The sum cannot in general be expressed in terms of a simple function. It can be expressed in terms of what are called beta functions but that is beyond the scope of this book. In most cases it should not be difficult to evaluate the sum for small values of m. The expectation and variance are given by:

$$
\begin{aligned}
E[X] &= \frac{r}{p} \qquad (3.35) \\
\mathrm{Var}[X] &= \frac{r(1-p)}{p^2}
\end{aligned}
$$

The other way to generalize things is to ask for the probability that k 0's occur before you get r 1's. This means that there are a total of $r+k$ tosses and the last toss gives you the rth 1. If X is the random variable and $r = 4$ then for the sequence 110001001 X would have the value 5. The probability that X has the value k is given by

$$P(X = k) = \binom{r+k-1}{k} p^r (1-p)^k \qquad (3.36)$$

This is also called a negative binomial distribution. It is just a slightly different version of the one defined above. If X_1 is the random variable for the first version, with probability defined in equation 3.33 and X_2 is the second version, with probability defined in equation 3.36 then the two are related as follows: $X_2 = X_1 - r$. For the probability that X is less than or equal to m, sum equation 3.36 from 0 to m. The expectation and variance are given by:

$$E[X] = \frac{r(1-p)}{p} \qquad (3.37)$$
$$\mathrm{Var}[X] = \frac{r(1-p)}{p^2}$$

3.7 Poisson Distribution

We mentioned that using a normal distribution to approximate the binomial works best when the probability of heads is close to $1/2$, $p \approx 1/2$. If p is close to 0 then the approximation will be poor. The reason is that the binomial distribution becomes very skewed for small p whereas the normal distribution is always symmetric about its mean. The way to approximate the binomial distribution for small p is to use the Poisson distribution. Let X be a Poisson random variable then the probability distribution is given by

$$P(X = k) = \frac{\lambda^k e^{-\lambda}}{k!} \qquad (3.38)$$

To use this to approximate a binomial distribution let $\lambda =$

np. The approximation will generally work well when $p \leq 0.1$ and $n \geq 10$. Both the expectation and variance are equal to λ. With $\lambda = np$ the expectation is the same as the binomial but the variance is slightly larger. Figure 3.8 shows a plot of both the binomial and Poisson distribution for $n = 20$, $p = 0.1$, and $\lambda = 2$. Table 3.1 shows the actual values. We only show k up to 10 because beyond that the values are very near zero. You can see that the agreement is generally good for low values of k but the binomial distribution falls off faster for larger values of k than the Poisson distribution. The reason is the binomial distribution is only defined for values of k that are less than or equal to n while the Poisson distribution is defined for all integer values of k from 0 to $+\infty$.

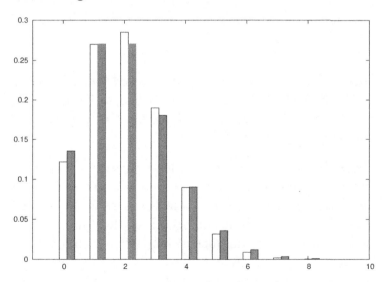

Figure 3.8: Binomial distribution (white) with $n = 20$, $p = 0.1$, and $k = 0, 1, \ldots, 10$ and its Poisson distribution approximation (gray) with $\lambda = np = 2$.

k	Binomial dist.	Poisson approx.
0	0.1215766546	0.1353352832
1	0.2701703435	0.2706705665
2	0.2851798071	0.2706705665
3	0.1901198714	0.1804470443
4	0.08977882815	0.09022352216
5	0.03192136112	0.03608940886
6	0.008867044756	0.01202980295
7	0.00197045439	0.003437086558
8	3.5577648711E-4	8.592716396E-4
9	5.2707627719E-5	1.9094925324E-4
10	6.4420433879E-6	3.8189850649E-5

Table 3.1: Binomial distribution with $n = 20$, $p = 0.1$, and $k = 0, 1, \ldots, 10$ and its Poisson distribution approximation with $\lambda = np = 2$.

The Poisson distribution was discovered by Simeon Denis Poisson (1781-1840) and published in his book *Research on the Probability of Judgments in Criminal and Civil Matters* (1837).

The Poisson distribution is generally used to model the probability that some random event occurs k times in some fixed interval of time. λ in this case is the average number of times the event occurs in the time interval. One example is the number of requests a web server receives per minute. Another example is the number of cars that cross some lonely rural intersection in an hour. In both cases the probability that the event occurs in any given instant of time can be vanishingly small but still add up to a finite average over the given time period.

Figure 3.9: Simeon Denis Poisson (1781-1840). Image credit: Francois Seraphin Delpech, Wikimedia.org

3.8   Exponential Distribution

Another continuous probability distribution related to coin tossing is the exponential distribution. It is essentially the continuous limit of the geometric distribution.

Let $F(n)$ be the probability that it takes less than or equal to n tosses to get a head then from the above section on the geometric distribution, by equation 3.30, we know that

$$F(n) = 1 - (1 - p)^n \qquad (3.39)$$

Now suppose that each toss takes time δt to complete. This means that if t is the time required for n tosses then $n = t/\delta t$. Define the rate at which heads appear as $\lambda = np/t$ then the probability of heads can be written as $p = \lambda\delta t$. Substituting these definitions into equation 3.39, $F(n)$ can be written as a function of time as follows

$$F(t) = 1 - (1 - \lambda\delta t)^{t/\delta t} \tag{3.40}$$

Now if you take the limit of this as δt goes to zero you get:

$$F(t) = 1 - e^{-\lambda t} \tag{3.41}$$

This is the probability that the first head occurs at some time less than t when the coin is being tossed continuously. At every instant of time the coin is tossed so that for any finite amount of time there are an infinite number of tosses and every toss takes a vanishingly small amount of time. Because of the way p was defined in terms of λ and δt it means that p must also be vanishingly small.

Equation 3.41 is the cumulative distribution function for the exponential probability distribution. To get the probability density function you simply take the derivative to get:

$$p(t) = \lambda e^{-\lambda t} \tag{3.42}$$

The exponential distribution occurs in physics in the length of time it takes for an excited atom to emit a photon or the time it takes for an unstable nucleus to decay. So you can look at this as the atom continuously tossing a coin to decide: should I or should I not emit a photon? For each toss the probability is very small but since the tossing is continuous it will eventually happen.

First Flight Centennial Commemorative Five Dollar Coin.
Image credit: United States Mint.

4 Betting on Coin Tosses

We will now look at the question of playing a betting game based on coin tossing. The game is simple. You guess what the next toss will be, heads (1) or tails (0). The coin is tossed and if your guess was correct you win $1. If it was incorrect you lose $1. A daily bet in the stock market is a simple variation of this game. If we ask the question: For each day in the past, did the stock market go up or down? and call a day where it was unchanged to be a down day, we get a series of YES and NO's that represents the daily history of the stock market. This series of YES and NO's could be interpreted as the results of flipping a coin, where heads equals YES, and tails equals NO. We should keep in mind that the coin might be biased, so that it might be far more likely to get heads than tails. And we should also keep in mind that we don't have to stick to a time period of a day; it could just as well be a minute, an hour, or a week. If you guess that the market will go up and it does then you will generally win more than $1. If your guess is wrong then you will generally lose more than $1. Other than the difference in the amounts won and lost, the dynamics are the same as a simple coin toss betting game.

First we'll describe the analysis we will do, and the conclusions we reach, and then show the details of the analysis, if you care to follow.

We start by looking at the case of a single biased coin where we know what the bias is. Finding the expectation for this, we reach the obvious conclusion that if we know

41

the direction of the constant bias, then we should always bet in that direction.

We then move on to the more common case of assuming there is a constant bias, but not knowing what the bias is, and asking the question for this case: What happens if I bet that today will be just like yesterday? We find that using this strategy ensures we will always be betting with the bias, as long as there is a bias. So we get a positive expectation in this case. We call this strategy BSP (Bet Same as Previous).

Is there a more optimal strategy than BSP when there is a constant bias but you don't know its direction? We look at an alternative to the BSP strategy, called the majority rule strategy. With this strategy, you count how many times in the past there has been an up day, and how many times there has been a down day, and then bet on whichever is larger. We show that when the history length is very large (technically infinity), then the expectation is the same as when you know the bias and you always bet in that direction. One question that must be answered to use the majority rule strategy is: How much historical data do you use? You could use as much as possible, or the last n days, for example.

4.1 Known Bias

To analyze the coin toss betting game we need to define some random variables. Let T_i represent the ith coin toss, and G_i represent the ith guess. Let's represent heads by the symbol '1' and tails by '0'. The probability of a coin toss coming up heads, we'll say is a constant p, and so the probability of getting tails is $1 - p$, because the two probabilities must add to 1. Writing these probabilities in mathematical form we have:

$$\begin{aligned} P(T_i = 1) &= p \\ P(T_i = 0) &= 1 - p \end{aligned} \qquad (4.1)$$

We now have to look at the question of how a guess is produced. To make things general, we will assume that the process of guessing is also random and completely independent of any of the toss values. Let the probability of guessing heads be q, and so the probability of guessing tails is $1 - q$. In mathematical form we have:

$$\begin{aligned} P(G_i = 1) &= q \\ P(G_i = 0) &= 1 - q \end{aligned} \qquad (4.2)$$

Winning the game means guessing correctly and losing means guessing wrong. So we win when T_i and G_i are both 1, or both 0. The probability of winning is then[1]

---

[1]Going from the 1st to the 2nd line of equation 4.3 assumes independence between the binary random processes of the coin toss and the guess. See appendix A for a definition of independence.

$$
\begin{aligned}
P_{\text{win}} &= P(T_i G_i = 11) + P(T_i G_i = 00) \qquad (4.3)\\
&= P(T_i = 1)P(G_i = 1) + P(T_i = 0)P(G_i = 0)\\
&= pq + (1 - p)(1 - q)\\
&= pq + 1 - p - q + pq = 2pq - p - q + 1
\end{aligned}
$$

And we lose when T_i and G_i are not the same. The probability of losing is then

$$
\begin{aligned}
P_{\text{lose}} &= P(T_i = 1)P(G_i = 0) + \qquad (4.4)\\
&\quad P(T_i = 0)P(G_i = 1)\\
&= p(1 - q) + (1 - p)q = p + q - 2pq
\end{aligned}
$$

To keep it simple, we'll say a win means you receive \$1, and a loss means you lose \$1. The expectation is the probability of winning, times the amount you win (+\$1), plus the probability of losing, times the amount you lose (-\$1), which is

$$
\begin{aligned}
E &= P_{\text{win}} - P_{\text{lose}} \qquad (4.5)\\
&= 4pq - 2p - 2q + 1 = (2p - 1)(2q - 1)
\end{aligned}
$$

It's natural to think of a probability as likely or unlikely, or in other words, how much bias it has. When a probability is $1/2$ we say there is no bias, so a bias shows how far the probability is from $1/2$. The probability of a coin toss coming up heads can be written in a form that shows the bias explicitly as follows:

$$p = \frac{1}{2} + a \qquad (4.6)$$

The parameter a is the bias. It can range in value from $-1/2$ to $+1/2$. If $a = 1/2$ then the probability of heads is certain, $p = 1$. If $a = -1/2$ then there is zero chance of getting heads, $p = 0$.

Similarly, the guessing probabilities we will write in bias form as:

$$q = \frac{1}{2} + b \qquad (4.7)$$

The parameter b is the guessing bias. If $b = 1/2$ then the guess will always be 1. If $b = -1/2$, the guess will always be 0.

Substituting equations 4.6 and 4.7 into 4.5, we get the expectation in terms of the biases a and b

$$E = (2a)(2b) = 4ab \qquad (4.8)$$

This shows the expectation is positive if both a and b have the same sign, that is, both the toss and the guess are biased in the same direction. For example, if both a and b are positive then both the toss and the guess are more likely to equal 1, and you get a positive expectation.

If you happen to know that the bias for the toss, a, is positive then it is optimal to make $b = +1/2$, that is to always guess 1, so $G_i = 1$ for all i. Then from equation 4.8, you get an expectation of $E = 2a$. If you don't know for sure that a is positive, then it's probably not a good idea to make $b = +1/2$ (always guess $G_i = 1$) because this could result in a maximum negative expectation if a turns out to be negative.

In most cases, you don't know for sure what the coin toss bias is, so you don't want to make your guesses constant. In the next section we look at what you can do in that case.

4.2   Unknown Bias

If you don't know the sign of the coin toss bias, a, then what can you do? There are two strategies that can be used in this case.

4.2.1   BSP Strategy

One strategy is to make the guess equal to the previous toss, that is $G_i = T_{i-1}$. This we call the bet the same as previous (BSP) strategy. What is the expectation for the BSP strategy?

The probability of winning with the BSP strategy is

$$
\begin{aligned}
P_{\text{win}} &= P(T_iG_i = 11) + P(T_iG_i = 00) && (4.9)\\
&= P(T_iT_{i-1} = 11) + P(T_iT_{i-1} = 00)\\
&= P(T_i = 1)P(T_{i-1} = 1) + P(T_i = 0)P(T_{i-1} = 0)\\
&= p^2 + (1 - p)^2\\
&= (\tfrac{1}{2} + a)^2 + (\tfrac{1}{2} - a)^2
\end{aligned}
$$

and the probability of losing with the BSP strategy is

$$
\begin{aligned}
P_{\text{lose}} &= P(T_iG_i = 10) + P(T_iG_i = 01) && (4.10)\\
&= P(T_iT_{i-1} = 10) + P(T_iT_{i-1} = 01)\\
&= P(T_i = 1)P(T_{i-1} = 0) + P(T_i = 0)P(T_{i-1} = 1)\\
&= p(1 - p) + (1 - p)p\\
&= 2(\tfrac{1}{2} + a)(\tfrac{1}{2} - a)
\end{aligned}
$$

These probabilities depend on the the coin tosses being independent of each other which is always true for the single coin model.

With a gain of $1 when we win and a loss of $1 when we lose, the expectation for the BSP strategy is:

$$
\begin{aligned}
E &= P_{\text{win}} - P_{\text{lose}} && (4.11)\\
&= (\frac{1}{2} + a)^2 + (\frac{1}{2} - a)^2 - 2(\frac{1}{2} + a)(\frac{1}{2} - a)\\
&= (\frac{1}{2} + a - (\frac{1}{2} - a))^2\\
&= (2a)^2 = 4a^2
\end{aligned}
$$

Comparing this expectation with the constant bias expectation of equation 4.8, you can see that the BSP strategy has the effect of ensuring that the guess bias is equal to the toss bias, that is $b = a$. This means that the expectation will always be positive with this strategy as long as $a \neq 0$.

4.2.2   Majority Rule Strategy

Is the BSP strategy the best you can do? Another strategy you can try when you don't know the sign of the coin toss bias is the majority rule strategy. In this strategy, you count the number of previous 1's and 0's, and whichever is larger is your guess for the next toss.

Assume there has been a $i = 0$ toss so that T_0 is known. Then the $i = 1$ guess will just use the BSP strategy because there is just 1 history point. So from equation 4.11, the expectation for the first toss is $4a^2$. In general, after an odd number of tosses, the expectation is (see equation 2.4 for a definition of binomial coefficients)

$$E_{2n+1} = (x-y)^2(1 + 2xy + 6x^2y^2 + 20x^3y^3 \qquad (4.12)$$
$$+70x^4y^4 + \ldots + \binom{2n}{n}x^ny^n)$$

where

$1, 2, 6, 20, 70, \ldots$ = central binomial coefficients

$$x = p = \frac{1}{2} + a, \quad y = 1 - p = \frac{1}{2} - a$$
$$x - y = 2a, \quad (x-y)^2 = (2a)^2 = 4a^2$$

and after an even number of tosses, the expectation is

$$E_{2n} = (x-y)^2(1 + \kappa_n(2xy + 6x^2y^2 \qquad (4.13)$$
$$+20x^3y^3 + 70x^4y^4 + \ldots + \binom{2n-2}{n-1}x^{n-1}y^{n-1}))$$

The expectations for even and odd are different because for an even number of tosses it is possible for there to be no majority in which case you have to resort to using the BSP strategy. This has the effect of reducing the expectation for the even number of tosses slightly. The factor κ_n in equation 4.13 shows the amount by which the expectation is reduced for an even number of tosses. The constant does not have a simple form and its derivation is beyond the scope of this book. It can be shown to converge to 1 as n increases so that $E_{2n} = E_{2n+1}$ for $n = \infty$. It can also be shown that when $n = \infty$ the power series in the equations converges to

$$1+2xy+6x^2y^2+20x^3y^3+70x^4y^4+\ldots = \frac{1}{x-y} \quad (4.14)$$

and the expectation is then

$$E_\infty = x - y = 2a \quad (4.15)$$

This is as large as the expectation can get. It is the same expectation you would get if you knew what the bias was and always bet in its direction.

In the case of an unknown bias, the conclusion is that you should use the majority rule strategy. This is based however on the premise that the bias remains constant, even though you don't know what it is. If the bias is changing, and especially if it has the potential for switching direction, then the BSP strategy is probably better since it can track a switching bias, as long as the switching is not too frequent.

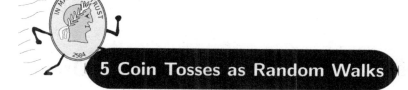
A series of coin tosses can be interpreted as a random walk on the integers with steps of size $+1$ or -1. You can picture this as a random walk on a line (x-axis) with hops between integer points. For example, if the walk starts at the integer k and you toss a head, you take a step to the right to $k + 1$. If you toss a tail, you take a step to the left to $k - 1$. The probability of a step to the right is p, the same as heads, and the probability of a step to the left is $q = 1 - p$, the same as tails. If $p = q = 1/2$ then the walk is symmetric, a step to the right is just as likely as a step to the left. This model can be very useful as a way of visualizing what happens in a series of coin tosses. It is particularly useful in analyzing gambling situations where you may lose or win some fixed amount on each game and you want to know the probability of losing all your money. This is the gambler's ruin problem which is discussed below.

We will start by showing how to count the number of walks between integer points on the line. This will be done in terms of generating functions for the number of walks. The generating functions will then be modified so that they become probability generating functions for walks between the points. The final section looks at the gambler's ruin problem.

5.1 Walks Returning to the Origin

To return to the origin there must be an even number of steps. Let the number of steps be $2n$ and let a_n be the number of walks of length $2n$. To get back to the origin there must be n steps to the left and n steps to the right, therefore a_n is given by the central binomial coefficients (EIS = A000984) [1].

$$a_n = \binom{2n}{n}$$

$$a_0 = 1$$

The recurrence equation for a_n is

$$na_n = 2(2n - 1)a_{n-1}$$

The first few values of a_n are shown in table 5.1.

n	0	1	2	3	4	5	6	7	8
a_n	1	2	6	20	70	252	924	3432	12870

Table 5.1: Number of walks returning to origin in $2n$ steps.

For example, for $n = 2$, the walk length is 4, and there are 6 such walks returning to the origin: +1+2+1, -1-2-1, +1+0-1, -1+0+1, +1+0+1, and -1+0-1.

---

[1] EIS refers to the Encyclopedia of Integer Sequences and A000984 is the sequence number.

The generating function for the a_n is

$$A(z) = \sum_{n=0}^{\infty} a_n z^{2n}$$
$$= \frac{1}{\sqrt{1 - 4z^2}}$$

Some fraction of the walks will be walks that are returning for the first time. Let f_n be the number of these walks. To get a relationship between the a_n and the f_n note that a walk can return to the origin after $2n$ steps by returning for the first time after 2 steps, and then returning finally after $2n - 2$ steps, or it can return for the first time after 4 steps and return finally after $2n - 4$ steps and so on. This means that a_n can be expressed as

$$a_n = \sum_{k=1}^{n} f_k a_{n-k}$$

and the 2 generating functions are related as follows

$$F(z) = 1 - \frac{1}{A(z)}$$
$$= 1 - \sqrt{1 - 4z^2}$$

Expanding $F(z)$ using the binomial theorem gives

$$F(z) = \sum_{n=1}^{\infty} f_n z^{2n}$$

The coefficients are

$$f_n = \frac{1}{2n-1}\binom{2n}{n}$$

The first two terms are $f_0 = 0$ and $f_1 = 2$. A recurrence equation for f_n is

$$n f_n = 2(2n-3) f_{n-1}$$

Table 5.2 shows the values of f_n up to $n = 8$.

n	1	2	3	4	5	6	7	8
f_n	2	2	4	10	28	84	264	858

Table 5.2: Number of walks of length $2n$ returning for first time.

The number of walks that return to the origin for the first time after $2n$ steps and always stay to the right of the origin is $f_n^+ = f_n/2$. The generating function is

$$F^+(z) = \frac{1}{2}(1 - \sqrt{1 - 4z^2})$$

The first few values are shown in table 5.3.

These are known as the Catalan numbers (EIS = A000108).

Let a_n^+ be the number of walks that return to the origin and always stay to the right, and let $A^+(z)$ be the generating function for these numbers, then

n	1	2	3	4	5	6	7	8
f_n^+	1	1	2	5	14	42	132	429

Table 5.3: Number of walks of length $2n$ returning for first time after $2n$ steps and always staying to the right of the origin.

$$A^+(z) = \frac{1}{1 - F^+(z)}$$
$$= \frac{1 - \sqrt{1 - 4z^2}}{2z^2}$$
$$= \sum_{n=0}^{\infty} a_n^+ z^{2n}$$
$$\text{where } a_n^+ = f_{n+1}^+$$

To get the probabilities associated with these walks, the generating functions only have to be modified slightly. Walks returning to the origin must have the same number of right and left steps (heads and tails) so each walk of length $2n$ will have a $p^n q^n$ probability. In the power series expansion of the generating functions the coefficient of z^{2n} must be this probability multiplied by the number of walks. The probability generating function for the return to the origin must then be

$$A(z) = \frac{1}{\sqrt{1 - 4pqz^2}}$$

and the probabilities are

$$a_n = \binom{2n}{n} p^n q^n$$

The probability generating function for a first return to the origin is

$$F(z) = 1 - \sqrt{1 - 4pqz^2}$$

and the probabilities are

$$f_n = \frac{1}{2n-1} \binom{2n}{n} p^n q^n$$

The probability that a walk returns to the origin at some point is the sum of all the first return probabilities. Call this probability P_0 then

$$P_0 = F(1) = 1 - \sqrt{1 - 4pq}$$

Using the fact that $1 - 4pq = (p - q)^2$, P_0 becomes

$$P_0 = 1 - |p - q|$$

When $p = q = 1/2$ then $P_0 = 1$ and a return to the origin is certain. For any other value of p there is a probability that a return never occurs. The probability of no return is

$$1 - P_0 = |p - q|$$

The probability that a walk is at the origin k times is $P_0^{k-1}(1 - P_0)$ (the start at the origin counts as 1 so it only has to return $k - 1$ times). Let V_0 be the number of times the walk visits the origin. The expected number of visits to the origin is then

$$
\begin{aligned}
E[V_0] &= \sum_{k=1}^{\infty} k P_0^{k-1}(1 - P_0) \\
&= (1 - P_0)\frac{d}{dP_0} \sum_{k=0}^{\infty} P_0^k \\
&= (1 - P_0)\frac{d}{dP_0} \left(\frac{1}{1 - P_0}\right) \\
&= \frac{1}{1 - P_0} \\
&= \frac{1}{|p - q|}
\end{aligned}
$$

For an unbiased walk (fair coin) where $p = q = 1/2$ this equation indicates that the walk will return to the origin an infinite number of times.

5.2   Walks from the Origin to m

A walk that starts at 0 and ends at $m > 0$ with no visit to m until the end is called a first visit walk. What are the number of first visit walks from 0 to 1? Note first that the length of the walk has to be an odd number, which can be expressed as $2n + 1$ with $n \geq 0$. For the walk to end at 1, it must be at 0 at step $2n$. The walk must stay on the negative integers or 0 until step $2n$. The number of these walks is the same as the number of walks that return to the origin and always stay to the right, which in the previous section we called a_n^+. The generating function for these numbers is $A^+(z)$. To add the extra step to get to 1 this generating function has to be multiplied by z to get

$$F_1(z) = zA^+(z)$$
$$= \frac{1 - \sqrt{1 - 4z^2}}{2z}$$

This is the generating function for the number of first visit walks from 0 to 1 of length $2n + 1$ which we will call $f_n^{(1)}$ so that

$$F_1(z) = \sum_{n=0}^{\infty} f_n^{(1)} z^{2n+1}$$

Let $a_n^{(1)}$ be the number of walks from 0 to 1 of length $2n + 1$ with no restrictions on the number of visits to 1. Such a walk can start at 0 and reach 1 for the first time at step $2n + 1$ or it can be at 0 at step 2 and then reach 1 for the first time in the remaining $2n - 1$ steps and so on. This can be expressed as

$$a_n^{(1)} = f_n^{(1)} a_0 + f_{n-1}^{(1)} a_1 + \cdots + f_0^{(1)} a_n$$

This is a convolution of the $f_n^{(1)}$ and a_n numbers, so the generating function for $a_n^{(1)}$ is a product of the generating functions for $f_n^{(1)}$ and a_n.

$$
\begin{aligned}
A_1(z) &= A(z)F_1(z) \\
&= \frac{1 - \sqrt{1 - 4z^2}}{2z} \frac{1}{\sqrt{1 - 4z^2}} \\
&= \sum_{n=0}^{\infty} a_n^{(1)} z^{2n+1}
\end{aligned}
$$

Using similar arguments to above it is possible to show that the generating function for the number of first visit walks from 0 to m of length $2n + m$ is given by

$$
\begin{aligned}
F_m(z) &= F_1^m(z) \\
&= \left(\frac{1 - \sqrt{1 - 4z^2}}{2z} \right)^m \\
&= \sum_{n=0}^{\infty} f_n^{(m)} z^{2n+m}
\end{aligned}
$$

The coefficient $f_n^{(m)}$ is equal to the number of first visit walks from 0 to m of length $2n + m$.

Likewise the generating function for $a_n^{(m)}$ is given by

$$A_m(z) = A(z)F_m(z) = A(z)F_1^m(z) = A_{m-1}(z)F_1(z)$$
$$= A(z)F_m(z) = \left(\frac{1 - \sqrt{1 - 4z^2}}{2z}\right)^m \frac{1}{\sqrt{1 - 4z^2}}$$
$$= \sum_{n=0}^{\infty} a_n^{(m)} z^{2n+m}$$

The coefficient $a_n^{(m)}$ is equal to the number of walks from 0 to m of length $2n + m$.

The following shows the expansion of the first four $F_m(z)$ functions.

$$F_1(z) = z + z^3 + 2\,z^5 + 5\,z^7 + 14\,z^9 + 42\,z^{11} + 132\,z^{13}$$
$$+ 429\,z^{15} + 1430\,z^{17} + \cdots$$

$$F_2(z) = z^2 + 2\,z^4 + 5\,z^6 + 14\,z^8 + 42\,z^{10} + 132\,z^{12} + 429\,z^{14}$$
$$+ 1430\,z^{16} + 4862\,z^{18} + \cdots$$

$$F_3(z) = z^3 + 3\,z^5 + 9\,z^7 + 28\,z^9 + 90\,z^{11} + 297\,z^{13} + 1001\,z^{15}$$
$$+ 3432\,z^{17} + 11934\,z^{19} + \cdots$$

$$F_4(z) = z^4 + 4\,z^6 + 14\,z^8 + 48\,z^{10} + 165\,z^{12} + 572\,z^{14}$$
$$+ 2002\,z^{16} + 7072\,z^{18} + 25194\,z^{20} + \cdots$$

The first ten coefficients of the first ten $F_m(z)$ functions are shown in table 5.4.

To get a formula for $a_n^{(m)}$ note that in the $2n + m$ steps, n steps are to the left and $n + m$ steps are to the right. The formula must then be

$$a_n^{(m)} = \binom{2n + m}{n}$$

a formula for $f_n^{(m)}$ can be derived from the generating function. We will leave out the details and just state the result as

$$f_n^{(m)} = \frac{m}{2n + m} \binom{2n + m}{n}$$

Now we look at probability generating functions for walks to m. The probability of taking the $n + m$ steps to the right and the n steps to the left that are required to get to m is $p^{n+m}q^n$. The number of lattice walks must be multiplied by this probability which means the generating functions given above need to be slightly modified. The probability generating function for a visit to m becomes

$$A_m(z) = \left(\frac{1 - \sqrt{1 - 4pqz^2}}{2qz} \right)^m \frac{1}{\sqrt{1 - 4pqz^2}}$$

and the probabilities are

m	n										
	0	1	2	3	4	5	6	7	8	9	10
1	1	1	2	5	14	42	132	429	1430	4862	16796
2	1	2	5	14	42	132	429	1430	4862	16796	58786
3	1	3	9	28	90	297	1001	3432	11934	41990	149226
4	1	4	14	48	165	572	2002	7072	25194	90440	326876
5	1	5	20	75	275	1001	3640	13260	48450	177650	653752
6	1	6	27	110	429	1638	6188	23256	87210	326876	1225785
7	1	7	35	154	637	2548	9996	38760	149226	572033	2187185
8	1	8	44	208	910	3808	15504	62016	245157	961400	3749460
9	1	9	54	273	1260	5508	23256	95931	389367	1562275	6216210
10	1	10	65	350	1700	7752	33915	144210	600875	2466750	10015005

Table 5.4: The first ten coefficients of the first ten $F_m(z)$ functions.

$$a_n^{(m)} = \binom{2n + m}{n} p^{n+m} q^n$$

The probability generating function for a first visit to m becomes

$$F_m(z) = \left(\frac{1 - \sqrt{1 - 4pqz^2}}{2qz} \right)^m$$

and the probabilities are

$$f_n^{(m)} = \frac{m}{2n + m} \binom{2n + m}{n} p^{n+m} q^n$$

The probability that a walk visits m at some point is the sum of all the first visit probabilities. Call this probability P_m then

$$P_m = F_m(1) = \left(\frac{1 - \sqrt{1 - 4pq}}{2q} \right)^m$$
$$= \left(\frac{1 - |p - q|}{2q} \right)^m$$

For an unbiased walk where $p = q = 1/2$ we get $P_m = 1$ which means a visit to m is certain to occur at some point no matter how large m.

To calculate the average number of visits to m define an indicator variable $I_n^{(m)}$ which is equal to 1 if a walk is at

m after $2n + m$ steps and is equal to 0 otherwise. Let V_m be the number of visits to m then

$$V_m = \sum_{n=0}^{\infty} I_n^{(m)}$$

The average number of visits is the expectation of V_m

$$E[V_m] = \sum_{n=0}^{\infty} E[I_n^{(m)}]$$
$$= \sum_{n=0}^{\infty} a_n^{(m)} = A_m(1)$$

where $A_m(1)$ is equal to

$$A_m(1) = \left(\frac{1 - \sqrt{1 - 4pq}}{2q} \right)^m \frac{1}{\sqrt{1 - 4pq}}$$
$$= \left(\frac{1 - |p - q|}{2q} \right)^m \frac{1}{|p - q|}$$

For an unbiased walk this becomes infinite, meaning that the walk will visit m an infinite number of times.

5.3   Gambler's Ruin

You want to play a game where you have a probability p of winning \$1 and a probability $q = 1 - p$ of losing \$1. Let

a be the amount of money you start with. You are playing against an opponent with b dollars. The game stops when either of you have no money left and the other has all $a+b$ dollars. Instead of playing against an opponent you can look at this as a strategy of stopping as soon as you win b dollars. In any case the process can be modeled as a walk in the integers that starts at a and stops if the walk gets to 0 or $a+b$. If you get to 0 you have lost every thing and are ruined. The question is, what is the probability of ruin given that you start with a dollars? Call this probability r_a and in general let r_k be the probability of ruin after you have been playing and your bankroll is now at k dollars. There are two things that can happen, you can win \$1 with probability p and your chance of ruin becomes r_{k+1} or you can lose \$1 with probability $q = 1 - p$ and your chance of ruin becomes r_{k-1} the following relationship must therefore hold

$$r_k = pr_{k+1} + qr_{k-1} \tag{5.1}$$

This is a homogeneous difference equation that can be solved by assuming a solution of the form z^k. Substituting this into the equation gives

$$pz^2 - z + q = 0 \tag{5.2}$$

Solving for z produces the two solutions $z = 1, q/p$. So r_k is a linear combination of these two solutions

$$r_k = A + B(q/p)^k \tag{5.3}$$

To solve for the constants A and B we use the boundary conditions $r_0 = 1$ and $r_N = 0$ where $N = a + b$. This gives the two equations $A + B = 1$ and $A + B(q/p)^N = 0$. Solving these equations for A and B with $q/p = \alpha$ gives

$$r_k = \frac{\alpha^k - \alpha^N}{1 - \alpha^N} \tag{5.4}$$

This solution holds for all $\alpha \neq 1$.

When $p = q = 1/2$ we have $\alpha = 1$ and the solution in equation 5.4 is meaningless. To get the correct solution you have to take the limit of equation 5.4 as α goes to 1. This gives

$$r_k = 1 - \frac{k}{N} \tag{5.5}$$

The probability of ruin in this case decreases linearly from $r_0 = 1$ to $r_N = 0$.

When $p < 1/2$ and $q > 1/2$ the game is unfavorable, $\alpha = q/p > 1$. When $p > 1/2$ and $q < 1/2$ the game is favorable, $\alpha < 1$. These two situations along with the case $\alpha = 1$ are shown in figure 5.1.

For $\alpha > 1$ the probability of ruin increases quickly as you move away from N while for $\alpha < 1$ the increase is slower.

Suppose you plan to play until you have \$100. How much do you have to start with so that your chance of ruin is no more than 0.5? If the game is fair, $\alpha = 1$, the obvious answer is \$50. For favorable or unfavorable games the

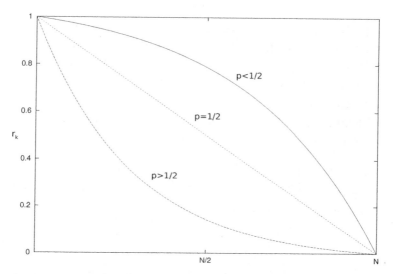

Figure 5.1: Probability of ruin with bankroll at k dollars for $p < 1/2$, $p > 1/2$ and $p = 1/2$.

amount is given by

$$k = \frac{\log(1 + \alpha^N) - \log(2)}{\log(\alpha)} \qquad (5.6)$$

For a game that is only slightly unfavorable with $p = 0.49$, $\alpha = 1.0408$ you will need at least \$83 for a 50% chance of winning \$17 (getting to \$100) before being ruined. On the other hand, with a slightly favorable game where $p = 0.51$ and $\alpha = 0.9608$ you only need \$17 for a 50% chance of winning \$83 without being ruined.

What happens if you play against an opponent with unlimited resources and you put no limits on when you will stop? To find out, take the limit as N goes to infinity of equation 5.4. There are two cases that have to be considered. In the favorable case where $p > q$ and $\alpha < 1$ the probability of ruin becomes $r_k = \alpha^k$. Here the chance of ruin decreases exponentially with increasing k. In the unfavorable case where $p < q$ and $\alpha > 1$ the probability of ruin is $r_k = 1$. The probability of ruin is certain in this case no matter how much money you start with.

On average how long will a game last until you either win or are ruined? Let N_k be the number of games played when starting from k then

$$N_k = \begin{cases} N_{k+1} + 1 & \text{with probability } p \\ N_{k-1} + 1 & \text{with probability } q \end{cases} \tag{5.7}$$

The expectation is then

$$E[N_k] = pE[N_{k+1} + 1] + qE[N_{k-1} + 1]$$
$$= pE[N_{k+1}] + qE[N_{k-1}] + 1$$

Let $c_k = E[N_k]$ then the equation can be written as an inhomogeneous difference equation

$$c_k - pc_{k+1} - qc_{k-1} = 1 \tag{5.8}$$

It is easy to verify that $c_k = k/(q - p)$ is a solution to this equation. Adding the solution to the corresponding homogeneous equation gives

$$c_k = A + B(q/p)^k + \frac{k}{q-p} \qquad (5.9)$$

The constants A and B are found from the boundary conditions $c_0 = 0, c_N = 0$ which give the equations $A + B = 0$, $A + B(q/p)^N + N/(q-p) = 0$. Solving for A and B and substituting into the above equation gives the solution

$$c_k = \frac{k}{q-p} - \frac{N}{q-p}\left(\frac{\alpha^k - 1}{\alpha^N - 1}\right) \qquad (5.10)$$

This solution is only valid when $\alpha \neq 1$ or $p \neq q$. When $p = q = 1/2$ the solution to equation 5.8 is $c_k = -k^2$ as can be readily verified. The general solution is then

$$c_k = A + Bk - k^2 \qquad (5.11)$$

Using the initial conditions and solving for A and B produces the solution

$$c_k = k(N - k) \qquad (5.12)$$

To find what happens against an opponent with unlimited resources take the limit as N goes to infinity. For $\alpha \leq 1$ or $q \leq p$, c_k becomes infinite meaning that the game could potentially go on forever. For $\alpha > 1$ or $q > p$ we have an unfavorable game and the average duration is

$$c_k = k/(q - p) \tag{5.13}$$

There are many more detailed questions which you can ask about the gambler's ruin problem. They can all be answered by knowing the generating function for the probability of a walk starting at k and ending at an arbitrary point j with absorbing nodes at 0 and N. The generating function is

$$G_{k,j}(z) = \begin{cases} \left(\frac{q}{p}\right)^{\frac{k-j}{2}} \dfrac{U_{j-1}\left(\frac{1}{2\sqrt{pqz}}\right) U_{N-k-1}\left(\frac{1}{2\sqrt{pqz}}\right)}{\sqrt{pqz} U_{N-1}\left(\frac{1}{2\sqrt{pqz}}\right)} & k \geq j \\[4ex] \left(\frac{p}{q}\right)^{\frac{j-k}{2}} \dfrac{U_{k-1}\left(\frac{1}{2\sqrt{pqz}}\right) U_{N-j-1}\left(\frac{1}{2\sqrt{pqz}}\right)}{\sqrt{pqz} U_{N-1}\left(\frac{1}{2\sqrt{pqz}}\right)} & k \leq j \end{cases} \tag{5.14}$$

where $U_n(x)$ is a type 2 Chebyshev polynomial. The first two Chebyshev polynomials are $U_0(x) = 1$, $U_1(x) = 2x$. The rest of them can be found from the following recurrence equation.

$$U_n(x) = 2x U_{n-1}(x) - U_{n-2}(x) \tag{5.15}$$

The recurrence can be solved to give the following closed form expression for $U_n(x)$.

$$U_n(x) = \frac{z_1^{n+1} - z_2^{n+1}}{z_1 - z_2} \tag{5.16}$$

where $z_1 = x + \sqrt{x^2 - 1}$ and $z_2 = x - \sqrt{x^2 - 1}$.

Deriving the generating function in 5.14 is a long and tedious process that would take us beyond the scope of this book. Its correctness can be easily verified. We leave this as an exercise and will now look at an example of how to use it.

Previously we calculated the probability of eventual ruin when starting at k. Using the generating function in 5.14 we can find the generating function for the probability that the ruin occurs on the nth step which we will call $R_k(z)$. Note that to get to 0, you first have to get to 1 and then take an additional step to the left. The probability generating function for getting to 1 from k is found by setting $j = 1$ in equation 5.14. The extra step to the left to 0 is represented by multiplying by qz. This gives the following expression for $R_k(z)$.

$$
\begin{aligned}
R_k(z) &= qzG_{k,1}(z) \\
&= \left(\frac{q}{p}\right)^{\frac{k}{2}} \frac{z_1^{N-k} - z_2^{N-k}}{z_1^N - z_2^N} \\
&= \left(\frac{q}{p}\right)^{k} \frac{\lambda_1^{N-k} - \lambda_2^{N-k}}{\lambda_1^N - \lambda_2^N}
\end{aligned}
$$

where

$$
z_1 = \frac{1 + \sqrt{1 - 4pqz^2}}{2\sqrt{pq}z}
$$

and

$$z_2 = \frac{1 - \sqrt{1 - 4pqz^2}}{2\sqrt{pq}z}$$

or in terms of the λ variables $\lambda_1 = \sqrt{q/p}z_1$ and $\lambda_2 = \sqrt{q/p}z_2$. The probability of ruin on the nth step is then equal to the coefficient of z^n in the power series expansion of $R_k(z)$. Our previous result for eventual ruin is found by evaluating $R_k(z)$ at $z = 1$, i.e. $r_k = R_k(1)$.

We can likewise find the generating function for the probability that a win occurs on the nth step (the walker is absorbed at N). Call this generating function $W_k(z)$. To get to N we first have to get $N - 1$ and then take an additional step to the right. The probability generating function for getting to $N - 1$ from k is found by setting $j = N - 1$ in equation 5.14. The extra step to the right is represented by multiplying by pz. $W_k(z)$ is then given by

$$
\begin{aligned}
W_k(z) &= pzG_{k,N-1}(z) \\
&= \left(\frac{p}{q}\right)^{\frac{N-k}{2}} \frac{z_1^k - z_2^k}{z_1^N - z_2^N} \\
&= \frac{\lambda_1^k - \lambda_2^k}{\lambda_1^N - \lambda_2^N}
\end{aligned}
$$

Let w_k be the probability of an eventual win (absorption at N), then $w_k = W_k(1)$

$$w_k = \frac{1 - (q/p)^k}{1 - (q/p)^N}$$

Note that $r_k + w_k = 1$ meaning that the walker is eventually absorbed at either 0 or N. To find the mean time to ruin, take the derivative of $R_k(z)$ and set $z = 1$. To find the mean time to a win, take the derivative of $W_k(z)$ and set $z = 1$. There are many more things you can do with these generating functions. We leave it to you to explore the possibilities.

Benjamin Franklin Commemorative Scientist Silver Dollar
Coin.
Image credit: United States Mint.

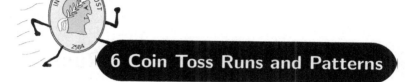

We are going to start by looking at runs as recurrent events. As an example of what this means, let's say we are interested in runs of length 3. In the sequence $THHH$ a run of 3 heads occurs on the fourth toss. At this point the run count starts over so that the fastest way to get another run of 3 heads is for the next 3 tosses all to be heads. In other words the runs cannot overlap. If the next toss is a head so that the sequence is $THHHH$, it does not count as another run of length 3. If the next 3 tosses are all heads so that the sequence is $THHHHHH$ then there are 2 runs of length 3 that occur on the fourth and seventh toss. If we are talking about runs of length 2 then there would be 3 in the sequence, occurring at the third, fifth, and seventh toss.

When runs are defined this way, they are recurrent events. When one occurs then every thing starts over as though we are starting a new sequence of tosses. A run is not the only possible recurrent event. Simple patterns such as the sequence $THTHTH$ can also be recurrent events with probabilities that can be calculated almost as easily as the simple run. A recurrent event can also be defined as not just a head run but as a head or a tail run. The reason for defining runs and other sequences as recurrent events is that it makes it much easier to find generating functions for probabilities and it generally simplifies the analysis. This will all become clear below.

6.1   Recurrence Times for Runs

Let's say you are watching a sequence of fair coin tosses at the rate of one per second and after about 40 minutes you suddenly see a run of 10 heads appear. Should you be surprised? Should you suspect that someone has found a way to tamper with the process? Maybe the coin has been switched and is no longer fair.

It's true that the probability of getting 10 heads in a row on a fair coin is pretty small at $1/1024 = 0.0009765625$. But even events with very small probabilities are likely to occur if you wait long enough. The question is how long do you have to wait, on average, to see such a run?[1]

The more general question is: given that a coin has a probability p of turning up heads on any single toss, how many times, on average, do you need to toss the coin to get a run of r heads? Table 6.1 shows this for 2-10 heads in a row given p from 0.1 to 0.9. Table 6.2 represents this information in terms of time with one toss per second. Note that the times vary from 2 seconds to 357 years. Table 6.3 is like table 6.1 but it shows the minimum number of tosses at which the probability of getting the run is greater than 0.5. Note that the toss counts of table 6.3 are all less than that of table 6.1, because the average number of tosses of table 6.1 have a minimum probability of 0.63 for getting the run. The QuantWolf Coin Toss Runs Calculator allows you to

---

[1]If you flip a fair coin once a second, the average waiting time for 10 heads in a row is 34 minutes, 6 seconds. So, yes you are likely to get 10 heads in a row in 40 minutes. See table 6.2.

calculate the average number of tosses needed to get at least r heads in a row, given the probability of heads.

Let f_n be the probability that it takes n tosses to get a run of r heads. Another way of saying this is that f_n is the probability that a run of length r occurs for the first time on the nth toss since the beginning or since the last time the run occurred. We will call f_n the recurrence time probability. If the f_n probabilities are known then the average number of tosses it takes to get a run of length r, called the mean recurrence time, is

$$\mu = \sum_{n=1}^{\infty} n f_n \qquad (6.1)$$

The problem then is to find the f_n values. Luckily it is easy to write down a recursion equation for f_n. To begin with, we know that $f_n = 0$ for $n < r$ and $f_r = p^r$. The latter being the case where the run appears in the first r tosses which has probability p^r. We also know that $f_n = p^r q$ for $r < n \leq 2r$. This is true because the last $r + 1$ tosses must be the r heads preceded by a tail which has probability $p^r q$. The first $n - r - 1$ tosses cannot possibly have a run of length r when $n \leq 2r$ and so can be ignored. These are the initial conditions. Now we just need a recursion equation to find the f_n values for $n > 2r$.

6.1.1 Recursion Equations for f_n

You can get a recursion equation for f_n by looking at the ways the sequence of tosses can start. With probability

r	p								
	0.1	0.2	0.3	0.4	0.5	0.6	0.7	0.8	0.9
2	110.00	30.00	14.44	8.75	6.00	4.44	3.47	2.81	2.35
3	1,110.00	155.00	51.48	24.37	14.00	9.07	6.38	4.77	3.72
4	11,110.00	780.00	174.94	63.44	30.00	16.79	10.55	7.21	5.24
5	111,110.00	3,905.00	586.46	161.09	62.00	29.65	16.50	10.26	6.94
6	1,111,110.00	19,530.00	1,958.20	405.23	126.00	51.08	25.00	14.07	8.82
7	11,111,110.00	97,655.00	6,530.68	1,015.59	254.00	86.81	37.14	18.84	10.91
8	111,111,110.00	488,280.00	21,772.26	2,541.46	510.00	146.34	54.49	24.80	13.23
9	1,111,111,110.00	2,441,405.00	72,577.52	6,356.16	1,022.00	245.57	79.27	32.25	15.81
10	11,111,111,110.00	12,207,030.00	241,928.40	15,892.91	2,046.00	410.95	114.67	41.57	18.68

Table 6.1: Average number of tosses needed to get at least r heads in a row. At the average, the probability of getting a run is a minimum of 0.63.

r				p					
	0.1	0.2	0.3	0.4	0.5	0.6	0.7	0.8	0.9
2	1m,50s	30s	14s	9s	6s	4s	3s	3s	2s
3	18m,30s	2m,35s	51s	24s	14s	9s	6s	5s	4s
4	3h,5m,10s	13m,0s	2m,55s	1m,3s	30s	17s	11s	7s	5s
5	1d,6h,51m,50s	1h,5m,5s	9m,46s	2m,41s	1m,2s	30s	16s	10s	7s
6	12d,20h,38m,30s	5h,25m,30s	32m,38s	6m,45s	2m,6s	51s	25s	14s	9s
7	4mn,8d,14h,25m,10s	1d,3h,7m,35s	1h,48m,51s	16m,56s	4m,14s	1m,27s	37s	19s	11s
8	3y,6mn,26d,0h,11m,50s	5d,15h,38m,0s	6h,2m,52s	42m,21s	8m,30s	2m,26s	54s	25s	13s
9	35y,8mn,20d,1h,58m,30s	28d,6h,10m,5s	20h,9m,38s	1h,45m,56s	17m,2s	4m,6s	1m,19s	32s	16s
10	357y,2mn,20d,19h,45m,10s	4mn,21d,6h,50m,30s	2d,19h,12m,8s	4h,24m,53s	34m,6s	6m,51s	1m,55s	42s	19s

Table 6.2: Average number of tosses, in terms of time (one toss per second), needed to get at least r heads in a row, where y=years, mn=months, d=days, h=hours, m=minutes, s=seconds

		p							
r	0.1	0.2	0.3	0.4	0.5	0.6	0.7	0.8	0.9
2	77	21	10	6	4	3	3	2	2
3	770	108	36	18	10	7	5	3	3
4	7,702	542	122	45	22	12	8	6	4
5	77,017	2,708	408	113	44	22	12	8	5
6	770,164	13,539	1,359	282	89	37	19	11	6
7	7,701,637	67,691	4,529	706	178	62	27	14	8
8	77,016,355	338,452	15,094	1,764	356	103	40	19	10
9	770,163,524	1,692,256	50,309	4,408	711	172	57	24	12
10	7,701,634,131	8,461,271	167,695	11,019	1,421	287	82	31	15

Table 6.3: Number of tosses at which the probability of getting at least r heads in a row is greater than or equal to 50%.

qf_{n-1}, the first toss is a tail and then the run occurs for the first time, $n-1$ tosses later. With probability pqf_{n-2} the first toss is a head followed by a tail and the run occurs for the first time $n-2$ tosses later. Continuing, you can get up to $r-1$ heads to start, followed by a tail with the run occurring $n-r$ tosses later. If you sum up all these probabilities you will then get the probability of the run occurring for the first time on the nth toss. This means the recursion equation must be:

$$f_n = qf_{n-1} + pqf_{n-2} + \cdots + p^{r-1}qf_{n-r} \qquad (6.2)$$
$$= q\sum_{k=1}^{r} p^{k-1} f_{n-k}$$

With the initial conditions discussed above, this equation can in principle be solved for the f_n which can then be used in equation 6.1 to find the mean recurrence time. But there is a better way to proceed and that is to find the generating function for the f_n. With the generating function you can easily calculate not just the mean but also the variance and higher moments. We will discuss the magic of generating functions below but first we look at a few more ways of calculating f_n and the corresponding cumulative distribution, F_n, which is equal to

$$F_n = \sum_{i=1}^{n} f_i \qquad (6.3)$$

The cumulative distribution is the probability that it takes less than or equal to n tosses to get a run of length r.

This is equivalent to the probability that a run occurs somewhere in the first n tosses. It is also equivalent to the probability that a run of length r or more occurs since you can't have a run of length greater than r without having a run of length r. Figure 6.1 shows the cumulative distribution for run lengths of 2 through 5 with a fair coin.

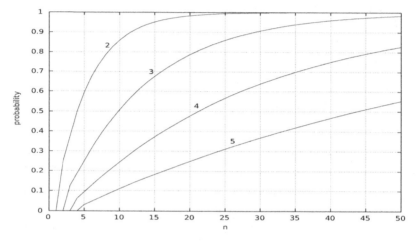

Figure 6.1: The probability that a run of length 2-5 heads occurs somewhere in the first n tosses of a fair coin.

F_n can be calculated from equation 6.3 but you can use some relationships between f_n and F_n to come up with more efficient ways of calculating both. To begin with we know that for a run to occur on toss n, there must be a tail on toss $n - r$ followed by r heads and this sequence has probability $p^r q$. For the run to be the first occurrence, it cannot have occurred on tosses 1 to $n - r - 1$ and this has probability:

$$1 - \sum_{i=1}^{n-r-1} f_i = 1 - F_{n-r-1} \qquad (6.4)$$

Another way of expressing f_n is therefore

$$f_n = p^r q(1 - F_{n-r-1}) \qquad (6.5)$$

Using this relationship for f_n, if you take the difference $f_n - f_{n-1}$ you get

$$\begin{aligned} f_n - f_{n-1} &= -p^r q(F_{n-r-1} - F_{n-r-2}) \qquad (6.6) \\ &= -p^r q f_{n-r-1} \end{aligned}$$

So we have another very simple recursion equation for f_n

$$f_n = f_{n-1} - p^r q f_{n-r-1} \qquad (6.7)$$

You can get a similar recursion equation for F_n from equation 6.5 by using $f_n = F_n - F_{n-1}$. This gives

$$F_n = F_{n-1} + p^r q(1 - F_{n-r-1}) \qquad (6.8)$$

To use this recursion you need the initial conditions for F_n. These can be found from the initial conditions for f_n. For $n < r$, $F_n = 0$, for $r \le n \le 2r$, $F_n = ((n-r)q+1)p^r$, and for $n > 2r$, equation 6.8 is used.

In addition to equations 6.2 and 6.7, there is a third way to calculate f_n. First write equation 6.2 in the following form

$$\frac{f_n}{p^n} = \frac{q}{p} \sum_{k=1}^{r} \frac{f_{n-k}}{p^{n-k}} \qquad (6.9)$$

Now define a new sequence $g_n = \frac{f_n}{p^n}$ and the equation becomes

$$g_n = \alpha \sum_{k=1}^{r} g_{n-k} \qquad (6.10)$$

where $\alpha = q/p = 1/p - 1$. The initial conditions for the equation are $g_1 = g_2 = \cdots = g_{r-1} = 0$, $g_r = 1$, $g_{r+1} = \alpha$. Calculating f_n by first calculating g_n is somewhat easier since equation 6.10 is a generalized Fibonacci type equation where the nth term is just proportional to the sum of the previous r terms in the sequence. These type of equations have been extensively studied and it is possible to find exact solutions for g_n when r is small. The equations will show up again when we look at probabilities for the longest run of heads on a fair coin.

6.1.2   Generating Functions for f_n

Now let's look at the generating function for f_n which we will call $F(z)$. If $G(z)$ is the generating function for g_n then $F(z)$ can be found by the simple variable substitution, $F(z) = G(pz)$. We will therefore find $G(z)$ first since it is somewhat easier. Start by multiplying equation 6.10 by z^n and summing over all n for which the equation is valid, that is from $n = r + 1$ to ∞. If you are familiar with z transforms then what we are doing is essentially taking the z transform of the equation (leaving aside questions of convergence). What we have then is

$$\sum_{n=r+1}^{\infty} g_n z^n = \alpha \sum_{n=r+1}^{\infty} (g_{n-1} + g_{n-2} + \cdots + g_{n-r}) z^n \quad (6.11)$$

The generating function we are looking for is defined as

$$\begin{aligned} G(z) &= \sum_{n=0}^{\infty} g_n z^n \quad\quad\quad (6.12) \\ &= \sum_{n=r}^{\infty} g_n z^n \\ &= z^r + \sum_{n=r+1}^{\infty} g_n z^n \end{aligned}$$

Comparing the left hand side of equation 6.11 with this definition you can see that it is equal to $G(z) - z^r$. On the right hand side we have terms of the form

$$\sum_{n=r+1}^{\infty} g_{n-k} z^n = \sum_{n=r+1-k}^{\infty} g_n z^{n+k} \qquad (6.13)$$

$$= z^k \sum_{n=r}^{\infty} g_n z^n$$

$$= z^k G(z)$$

Putting it all together, equation 6.11 becomes

$$G(z) - z^r = \alpha(z + z^2 + z^3 + \cdots + z^r) G(z) \qquad (6.14)$$

which you can then solve for $G(z)$ to get

$$G(z) = \frac{z^r}{1 - \alpha(z + z^2 + z^3 + \cdots + z^r)} \qquad (6.15)$$

This can be simplified somewhat by collapsing the geo-metric series in the denominator, giving

$$G(z) = \frac{z^r(1-z)}{1 - (\alpha+1)z + \alpha z^{r+1}} \qquad (6.16)$$

The generating function for f_n is then

$$F(z) = G(pz) \qquad (6.17)$$

$$= \frac{p^r z^r(1 - pz)}{1 - z + qp^r z^{r+1}}$$

6.1.3 Calculating Recurrence Time Statistics

The generating function for f_n can now be used to easily calculate recurrence time statistics. To see how $F(z)$ can be used to calculate the mean recurrence time in equation 6.1, write out $F(z)$ in the form of a power series:

$$F(z) = \sum_{n=0}^{\infty} f_n z^n \qquad (6.18)$$

Its derivative is then

$$F'(z) = \sum_{n=1}^{\infty} n f_n z^{n-1} \qquad (6.19)$$

Comparing this with equation 6.1 you can see that $\mu = F'(1)$. So if you take the derivative of $F(z)$ in equation 6.17 and set $z = 1$ you get

$$\mu = \frac{1 - p^r}{qp^r} \qquad (6.20)$$
$$= 1/p + (1/p)^2 + (1/p)^3 + \cdots + (1/p)^r$$

This is much easier than trying to calculate the infinite series form of the equation. An additional advantage of finding the generating function is that we can now easily calculate the variance also. The variance is defined as

$$\sigma^2 = \sum_{n=1}^{\infty} (n - \mu)^2 f_n \qquad (6.21)$$

$$= \sum_{n=1}^{\infty} n^2 f_n - \mu^2$$

To calculate this all we need is the second derivative of $F(z)$

$$F''(z) = \sum_{n=2}^{\infty} n(n-1) f_n z^{n-2} \qquad (6.22)$$

which evaluated at $z = 1$ is

$$F''(1) = \sum_{n=2}^{\infty} n^2 f_n - \mu \qquad (6.23)$$

The variance is then $\sigma^2 = \mu(1 - \mu) + F''(1)$ which if we evaluate the derivative and set $z = 1$ gives

$$\sigma^2 = \frac{1 - p^{2r+1} - qp^r(2r+1)}{q^2 p^{2r}} \qquad (6.24)$$

$$= \frac{1}{(qp^r)^2} - \frac{2r+1}{qp^r} - \frac{p}{q^2}$$

Higher moments can also be found in a similar manner.

6.1.4 Examples

Now we look at a couple of examples. The simplest case is where $r = 1$ so that f_n is the probability that a head appears for the first time on the nth toss. The first $n - 1$ tosses must be tails followed by a head, therefore $f_n = pq^{n-1}$ which is the well known geometric distribution. This is a trivial example where no recursion equation or generating function is needed. Nevertheless, let's look at this using the machinery developed above to see how it works. The recursion is just one term, $f_n = qf_{n-1}$, $f_1 = p$, and this obviously has the solution: $f_n = pq^{n-1}$. The generating function is

$$
\begin{aligned}
F(z) &= \frac{pz(1 - pz)}{1 - z + qpz^2} \\
&= \frac{pz}{1 - qz} \\
&= pz(1 + qz + q^2z^2 + q^3z^3 + \cdots)
\end{aligned}
\tag{6.25}
$$

The coefficient of z^n is $f_n = pq^{n-1}$ as it should be.

The next example is for $r = 2$ where f_n is now the probability that two heads in a row occur for the first time on the nth toss. The first $n - 2$ tosses can have both heads and tails as long as no two heads appear in a row and there is the restriction that toss $n - 2$ cannot be a head as this would cause the run to appear at $n - 1$ and not n. The situation is now more complicated and either the recursion equation or the generating function is needed to find f_n. This is a case where the generating function is still relatively simple and exact formulas can be found. Instead of finding f_n directly it is easier to find g_n and then use $f_n = p^n g_n$. The generating function for g_n is

$$G(z) = \frac{z^2}{1 - \alpha z - \alpha z^2} \qquad (6.26)$$

To get a formula for g_n from this, we do a partial fraction expansion of $G(z)$ so that we can write it as

$$G(z) = z_1 z_2 - \frac{A z_2}{1 - z/z_1} + \frac{B z_1}{1 - z/z_2} \qquad (6.27)$$

$$z_1 = -\frac{1}{2}(1 + \sqrt{1 + 4/\alpha})$$

$$z_2 = -\frac{1}{2}(1 - \sqrt{1 + 4/\alpha})$$

$$A = \frac{z_1^2}{z_1 - z_2}$$

$$B = \frac{z_2^2}{z_1 - z_2}$$

The two terms in z are then easily expanded as

$$\frac{A z_2}{1 - z/z_1} = A z_2 (1 + z/z_1 + (z/z_1)^2 + \cdots) \qquad (6.28)$$

$$\frac{B z_1}{1 - z/z_2} = B z_1 (1 + z/z_2 + (z/z_2)^2 + \cdots)$$

Collecting terms, you find that the coefficient of z^n is

$$g_n = \frac{\alpha^{n-2}}{2^{n-1}} \frac{(1 + \sqrt{1 + 4/\alpha})^{n-1} - (1 - \sqrt{1 + 4/\alpha})^{n-1}}{\sqrt{1 + 4/\alpha}}$$

$$(6.29)$$

Table 6.4 shows g_n for n up to 16 and probabilities of the form $p = 1/k$ where k is an integer from 2 to 6. In this case $\alpha = k - 1$, g_n is always an integer, and $f_n = g_n/k^n$. With f_n equal to an integer divided by k^n it suggests that what we have here is equivalent to a problem involving k possible outcomes or a k sided coin instead of the regular two sided heads or tails. For example when $p = 1/4$, the problem is equivalent to tossing a 4 sided coin with all sides having equal probability and f_n is the probability of getting a run of length 2, for one of the 4 sides for the first time on the nth toss. With this interpretation, g_n counts the number of strings of length n that are composed of 4 symbols with one particular symbol only appearing double at the end of the string. This maps onto the binary coin toss by making heads equal to one of the 4 symbols and tails equal to any of the remaining 3 symbols.

6.1.5 Approximations and Other Methods

In equation 6.29 the term $(1 + \sqrt{1 + 4/\alpha})^{n-1}$ will dominate the term $(1 - \sqrt{1 + 4/\alpha})^{n-1}$ for large n. This sug-

		$1/2$	$1/3$	p $1/4$	$1/5$	$1/6$
	2	1	1	1	1	1
n	3	1	2	3	4	5
	4	2	6	12	20	30
	5	3	16	45	96	175
	6	5	44	171	464	1025
	7	8	120	648	2240	6000
	8	13	328	2457	10816	35125
	9	21	896	9315	52224	205625
	10	34	2448	35316	252160	1203750
	11	55	6688	133893	1217536	7046875
	12	89	18272	507627	5878784	41253125
	13	144	49920	1924560	28385280	241500000
	14	233	136384	7296561	137056256	1413765625
	15	377	372608	27663363	661766144	8276328125
	16	610	1017984	104879772	3195289600	48450468750

Table 6.4: Values of g_n for runs of length 2.

gests that a good approximation for g_n for large n is

$$g_n^* = \frac{\alpha^{n-2}}{2^{n-1}} \frac{(1 + \sqrt{1 + 4/\alpha})^{n-1}}{\sqrt{1 + 4/\alpha}} \qquad (6.30)$$

Table 6.5 shows this approximation for $p = 1/3$, $\alpha = 2$. You can see that the approximation is actually quite good even for small values of n. You can generally find approximations like this for any value of r and p or α. The general idea is that in the partial fraction expansion of $G(z)$, the term for the smallest magnitude root of the denominator of $G(z)$ will dominate for large values of n and the others can be ignored in calculating g_n. In general if z_1 is the smallest magnitude root of $1 - \alpha(z + z^2 + z^3 + \cdots + z^r)$ and it is a simple root, then when n is large, a good approximation for g_n is

$$g_n \approx \frac{a_1/\alpha}{z_1^{n+1}} \qquad (6.31)$$

where a_1 is given by

$$a_1 = \frac{z_1^r}{1 + 2z_1 + 3z_1^2 + 4z_1^3 + \cdots + r z_1^{r-1}} \qquad (6.32)$$

We should also mention that it is possible to get the g_n values by differentiating $G(z)$. You can see this by looking at the way $G(z)$ is defined in equation 6.12. If you differentiate $G(z)$ n times and evaluate the result at 0

n	g_n	g_n^*
2	1	0.789
3	2	2.155
4	6	5.887
5	16	16.083
6	44	43.939
7	120	120.044
8	328	327.967
9	896	896.024
10	2448	2447.983
11	6688	6688.013
12	18272	18271.991
13	49920	49920.007
14	136384	136383.995
15	372608	372608.004
16	1017984	1017983.997

Table 6.5: Values of g_n and its approximation for runs of length 2 with $p=1/3$.

then you get $n!g_n$. So another formula for g_n is:

$$g_n = \frac{G^{(n)}(0)}{n!} \tag{6.33}$$

The differentiation must of course be done on the rational form of $G(z)$ given in equations 6.15 or 6.16. Calculating the nth derivative can become very complicated for large n and the only reasonable way to do it is with a computer algebra system such as Mathematica, Maple, or Maxima.

6.2 Multiple Recurrence Times for Runs

Now we want to look at the probability that it takes n tosses to get more than one run of r heads. Call this probability $f_n^{(k)}$ where k is the number of runs. When $k = 1$ we have $f_n^{(1)} = f_n$ which is what we have looked at so far. When $k = 2$ we want the probability that the second run of r heads occurs on the nth toss. To get a recursion equation for $f_n^{(2)}$ we need to take into account all the ways the second run can occur on toss n. We can have the first run at toss 1 and the second $n - 1$ tosses later or the first run at toss 2 and the second $n - 2$ tosses later and so on. The probability must therefore be

$$f_n^{(2)} = f_1 f_{n-1} + f_2 f_{n-2} + \cdots + f_{n-1} f_1 \tag{6.34}$$

The right hand side is just the convolution of the sequence

f_n with itself. The generating function for the convolution of two sequences is equal to the product of their generating functions. The generating function for $f_n^{(2)}$ is therefore

$$F^{(2)}(z) = \sum_{n=1}^{\infty} f_n^{(2)} z^n \qquad (6.35)$$
$$= F^2(z)$$

where $F(z)$ is the generating function for f_n given in equation 6.17. Note that the $n = 0$ was left out of the sum because $f_0^{(k)} = 0$ for all k. Using similar arguments you can see that $f_n^{(3)}$ is the convolution of f_n and $f_n^{(2)}$ so that the generating function is $F^{(3)}(z) = F^3(z)$. The generating function for $f_n^{(k)}$ is then

$$F^{(k)}(z) = \sum_{n=1}^{\infty} f_n^{(k)} z^n \qquad (6.36)$$
$$= F^k(z)$$

A simple example is for $r = 1$. We showed earlier that $f_n = pq^{n-1}$ in this case which is just the geometric distribution. Using the generating function or through repeated convolutions it is not hard to show that in general for $r = 1$

$$f_n^{(k)} = \binom{n-1}{k-1} p^k q^{n-k} \qquad (6.37)$$

which is the well known negative binomial distribution for the probability that in a sequence of n tosses the kth head appears on the nth toss.

Just like f_n can be expressed in terms of the simpler sequence $g_n = f_n/p^n$ so $f_n^{(k)}$ can be expressed in terms of the sequence $g_n^{(k)} = f_n^{(k)}/p^n$. The generating function for $g_n^{(k)}$ is then

$$G^{(k)}(z) = \sum_{n=1}^{\infty} g_n^{(k)} z^n \qquad (6.38)$$
$$= G^k(z)$$

where $G(z)$ is the generating function for g_n given in equation 6.15 or 6.16.

The mean and standard deviation of the multiple recurrence times can easily be calculated given the definition of the generating function in equation 6.36. They are given by

$$\mu^{(k)} = k\mu \qquad (6.39)$$
$$(\sigma^{(k)})^2 = k\sigma^2$$

where μ and σ^2 are the mean and standard deviation of the single recurrence time.

6.3 Run Occurrence Probability

What is the probability that a run occurs on toss n? The f_n values that we have been looking at give the probability that a run occurs for the first time at toss n. Now we simply want the probability that it occurs at toss n, not necessarily for the first time. Let a_n be the probability that a run of length r appears on the nth toss then we can get a recursion equation for a_n by looking at all the ways you can have r heads in a row at toss n. If the run occurs at toss n then by definition there are r heads in a row and this has probability a_n. If the run occurs at toss $n-1$ followed by a head then there will also be r heads in a row and this has probability pa_{n-1}. Continuing on like this you can go up to having the run occur at toss $n-r+1$ followed by $r-1$ heads to have r heads in a row at toss n. Summing up all the possibilities gives the following equation

$$a_n + pa_{n-1} + p^2 a_{n-2} + \cdots + p^{r-1} a_{n-r+1} = p^r \quad (6.40)$$

This equation is of course only valid for n greater than or equal to r with the initial conditions being $a_n = 0$ for $n < r$. To find the generating function just multiply both sides of the equation by z^n and sum from $n = r$ to ∞ (note that a_0 is defined as being equal to 1 here). The generating function is then

$$A(z) = \frac{1 - z + qp^r z^{r+1}}{(1-z)(1-p^r z^r)} \quad (6.41)$$

The a_n probabilities can be related to the f_n probabilities by looking at all the ways a run can occur on the nth toss. It can occur for the first time on the nth toss with probability f_n. It can occur at toss 1 and not reoccur until $n - 1$ tosses later with probability $f_{n-1}a_1$ or it can occur at toss 2 and not reoccur until $n - 2$ tosses later with probability $f_{n-2}a_2$ and so on. Summing up these probabilities gives

$$a_n = f_n a_0 + f_{n-1}a_1 + f_{n-2}a_2 + \cdots + f_1 a_{n-1} \quad (6.42)$$

This equation is valid for $n \geq 1$. Multiply both sides by z^n and sum from $n = 1$ to ∞ and you get

$$A(z) - 1 = F(z)A(z) \quad (6.43)$$

So the two generating functions are related to each other as follows

$$A(z) = \frac{1}{1 - F(z)} \quad (6.44)$$

$$F(z) = \frac{A(z) - 1}{A(z)} \quad (6.45)$$

Equation 6.44 could have been anticipated since when a run occurs at toss n, it is either the first, second, third or higher occurrence, therefore a_n must be equal to

$$a_n = f_n + f_n^{(2)} + f_n^{(3)} + \cdots \quad (6.46)$$

This means the generating function for a_n must be equal to

$$A(z) = 1 + F(z) + F^2(z) + F^3(z) + \cdots \qquad (6.47)$$

which is just the expansion of equation 6.44.

6.3.1 Examples

Let's look at the trivial case first where $r = 1$ so that a_n is just the probability that toss n results in heads. Immediately we know then that $a_n = p$ and all we have to do is verify that the generating function gives the same result. From equation 6.41 with $r = 1$ we get

$$
\begin{aligned}
A(z) &= \frac{1 - z + qpz^2}{(1 - z)(1 - pz)} \qquad (6.48) \\
&= \frac{1 - qz}{1 - z} \\
&= (1 - qz)(1 + z + z^2 + z^3 + \cdots) \\
&= 1 + pz + pz^2 + pz^3 + \cdots
\end{aligned}
$$

verifying that $a_n = p$.

For $r = 2$ we want the probability that toss n gives us two heads in a row. The following probabilities are immediately obvious: $a_1 = 0$, $a_2 = p^2$, and $a_3 = qp^2$. For the others we will use the generating function which in this case is

$$A(z) = \frac{1 - qz - pqz^2}{(1 - z)(1 + pz)} \tag{6.49}$$

$$= \frac{p^2}{1 + p}\frac{1}{1 - z} + \frac{p}{1 + p}\frac{1}{1 + pz} + q$$

Expanding the two partial fractions and collecting terms gives

$$a_4 = \frac{p^2(1 + p^3)}{1 + p} \tag{6.50}$$

$$a_5 = \frac{p^2(1 - p^4)}{1 + p} \tag{6.51}$$

$$a_n = \frac{p^2(1 + (-1)^n p^{n-1})}{1 + p} \tag{6.52}$$

The same procedure can be used to find a_n for any value of r. To get the solution for any r, let $k = n \bmod r$ then [2]

$$\mu a_n = \begin{cases} 1 + p^{n-r+1}(1 - p^{r-1})/(1 - p) & k = 0 \\ 1 - p^{n-k} & k \neq 0 \end{cases} \tag{6.53}$$

where μ is the mean recurrence time given in equation 6.20.

---

[2] mod here refers to modular arithmetic.

6.4   Head or Tail Run Probabilities

We now want to look at the probability that a run of heads or tails occurs. Let h be the length of the head run and t be the length of the tail run. Head and tail probabilities are, as before, p and $q = 1 - p$. The generating function for the head run occurrence probabilities are

$$A(z) = \frac{1 - z + qp^h z^{h+1}}{(1 - z)(1 - p^h z^h)} \qquad (6.54)$$

$$= \sum_{n=0}^{\infty} a_n z^n$$

and the generating function for the tail run occurrence probabilities are

$$B(z) = \frac{1 - z + pq^t z^{t+1}}{(1 - z)(1 - q^t z^t)} \qquad (6.55)$$

$$= \sum_{n=0}^{\infty} b_n z^n$$

Let $C(z)$ be the generating function for the head or tail run occurrence probabilities then

$$C(z) = A(z) + B(z) - 1 \qquad (6.56)$$
$$= \frac{1 - z + p^h q z^{h+1} + p q^t z^{t+1} - p^h q^t z^{h+t}}{(1 - z)(1 - p^h z^h)(1 - q^t z^t)}$$
$$= \sum_{n=0}^{\infty} c_n z^n$$

This comes from the fact that the probability of one or the other run occurring is just the sum of their individual probabilities so that $c_n = a_n + b_n$. We define $a_0 = b_0 = c_0 = 1$ which is why 1 has to be subtracted in equation 6.56.

The generating function for head or tail run recurrence time probabilities is then

$$F(z) = \frac{C(z) - 1}{C(z)} \qquad (6.57)$$
$$= \frac{p^h z^h (1 - pz)(1 - q^t z^t) + q^t z^t (1 - qz)(1 - p^h z^h)}{1 - z + p^h q z^{h+1} + p q^t z^{t+1} - p^h q^t z^{h+t}}$$
$$= \sum_{n=0}^{\infty} f_n z^n$$

To get the mean recurrence time you take the derivative of this with respect to z and evaluate at $z = 1$. This gives

$$\mu = F'(1) \qquad (6.58)$$
$$= \frac{(1 - p^h)(1 - q^t)}{p^h q + pq^t - p^h q^t}$$

Note that when $t = \infty$ then μ is just the mean recurrence time for a run of heads of length h and when $h = \infty$ then μ is the mean recurrence time for a run of tails of length t. The QuantWolf Coin Toss Runs Calculator allows you to calculate the average number of tosses (mean recurrence time) needed to get either h heads in a row or t tails in a row. Table 6.6 shows the average number of tosses needed to get either r heads or r tails in a row for probability p of tossing a head from 0.1 to 0.9. Note that as p approaches 1, the average number of tosses converges to the average number of tosses for just heads (compare the column of $p = 0.9$ to that of table 6.1).

The variance of the recurrence times is $\sigma^2 = \mu(1 - \mu) + F''(1)$. A general equation for this is too complicated to be of much use. The value of $F''(1)$ is probably best evaluated numerically in the general case. As examples, for the case of $h = t = 2$ we have:

$$F''(1) = \frac{2(1 + 5pq)}{(1 - pq)^2} \qquad (6.59)$$

and for $h = t = 3$ we have:

$$F''(1) = \frac{2(3 + 10pq + 11p^2q^2 - 4p^3q^3)}{(1 - 2pq - p^2q^2)^2} \qquad (6.60)$$

r	p								
	0.1	0.2	0.3	0.4	0.5	0.6	0.7	0.8	0.9
2	2.30	2.57	2.80	2.95	3.00	2.95	2.80	2.57	2.30
3	3.71	4.62	5.68	6.61	7.00	6.61	5.68	4.62	3.71
4	5.24	7.14	9.95	13.28	15.00	13.28	9.95	7.14	5.24
5	6.93	10.23	16.05	25.04	31.00	25.04	16.05	10.23	6.93
6	8.82	14.06	24.68	45.36	63.00	45.36	24.68	14.06	8.82
7	10.91	18.84	36.93	79.97	127.00	79.97	36.93	18.84	10.91
8	13.23	24.80	54.35	138.38	255.00	138.38	54.35	24.80	13.23
9	15.81	32.25	79.18	236.44	511.00	236.44	79.18	32.25	15.81
10	18.68	41.57	114.62	400.60	1,023.00	400.60	114.62	41.57	18.68

Table 6.6: Average number of tosses needed to get either r heads or r tails in a row for probability p of tossing a head from 0.1 to 0.9.

The recursion equation for f_n, the probability that a run of h heads or a run of t tails occurs for the first time on toss n, is somewhat messy to derive. We will simply list them below for the special cases of $h = t = r$ and $r = 2, 3, 4, 5, 6$. In the section on Markov models for runs and patterns you will see how to get f_n from the powers of a Markov matrix.

For $h = t = r$, the initial conditions for f_n are

$$f_1 = f_2 = \cdots = f_{r-1} = 0 \qquad (6.61)$$
$$f_r = p^r + q^r$$
$$f_{r+1} = f_{r+2} = \cdots = f_{2r-1} = p^r q + pq^r$$

The following recursion equations are then valid for $n \geq r$

r	$f_n =$
2	$pq f_{n-2}$
3	$pq(f_{n-2} + f_{n-3}) + (pq)^2 f_{n-4}$
4	$pq(f_{n-2} + f_{n-3}) + pq(1 - pq)f_{n-4} + (pq)^2 f_{n-5} + (pq)^3 f_{n-6}$
5	$pq(f_{n-2} + f_{n-3}) + pq(1 - pq)f_{n-4} + pq(1 - 2pq)f_{n-5} + (pq)^2(1 - pq)f_{n-6} + (pq)^3 f_{n-7} + (pq)^4 f_{n-8}$
6	$pq(f_{n-2} + f_{n-3}) + pq(1 - pq)f_{n-4} + pq(1 - 2pq)f_{n-5} + pq(1 - 3pq + (pq)^2)f_{n-6} + (pq)^2(1 - 2pq)f_{n-7} + (pq)^3(1 - pq)f_{n-8} + (pq)^4 f_{n-9} + (pq)^5 f_{n-10}$

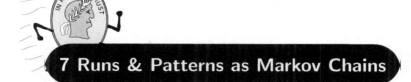

When runs and patterns are defined as recurrent events (see previous chapter for definition) then they can be modeled as Markov chains. A Markov chain is a set of states and a set of probabilities for transitions between the states. The basic idea is that any run or pattern must be built up from consecutive tosses until it is complete and each step in this process corresponds to a state of the Markov Chain.

The theory of Markov chains is powerful and well developed and it can allow you to analyze runs and patterns in a depth that is probably not possible in any other way. We are going to look at three examples of how to use Markov chains: head runs, head or tail runs, and simple patterns. Only the basic ideas will be illustrated since a detailed analysis would require a lengthy detour into Markov chain theory that is beyond the scope of this book.

7.1 Head Run Markov Chain

The simplest way to see how the Markov chain model works is to look at recurrent head runs of fixed length r. The states correspond to the number of consecutive heads that have occurred in the toss sequence and they are labeled accordingly s_0 to s_r. The toss sequence starts in state s_0. If the first toss is a head we move to state s_1. This means the transition from s_0 to s_1 happens with probability p. If the first toss is a tail then we stay in s_0.

The probability of staying in s_0 is therefore $q = 1 - p$. When in state s_1 we move to s_2 with probability p and back to s_0 with probability q. In general we move from s_k to s_{k+1} with probability p and we move from s_k to s_0 with probability q. The exception is when we get to s_r at which point a run has occurred and we stay in that state. In the language of Markov chain theory s_r is called an absorbing state. Figure 7.1 shows a state transition diagram for runs of length 3.

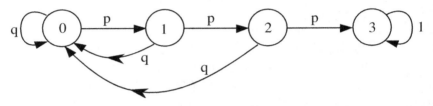

Figure 7.1: Recurrence time probability Markov chain for runs of length 3.

The Markov matrix for the chain is:

$$\mathbf{M} = \begin{pmatrix} q & p & 0 & 0 \\ q & 0 & p & 0 \\ q & 0 & 0 & p \\ 0 & 0 & 0 & 1 \end{pmatrix} \tag{7.1}$$

The matrix element $\mathbf{M}_{ij}$ is the probability of moving from state s_i to s_j in one toss.[1] The probability of moving from s_i to s_j in n tosses is given by $\mathbf{M}_{ij}^n$ where $\mathbf{M}^n$ is the nth power of the matrix. Since we always start out in s_0, only

---

[1] The rows and columns of the matrix are numbered from 0 to 3 and the ij subscript denotes the element at row i and column j.

the matrix elements in the first row will be of interest, that is the elements $\mathbf{M}_{0j}^{n}$. In particular for the $r = 3$ example above we want the probability, f_n, that a run occurs for the first time on toss n. For this to happen we have to be in state s_2 at toss $n - 1$ and the next toss has to be a head. The probability of this happening is $f_n = \mathbf{M}_{02}^{n-1} p$.

If we want to calculate the probability that a run occurs on toss n, denoted previously as u_n, then we have to use a slightly different Markov chain. Now the s_3 state can no longer be absorbing and we have to allow transitions out of it to s_0 if the next toss is a tail, or to s_1 if the next toss is another head. The state transition diagram and the new Markov matrix are shown below.

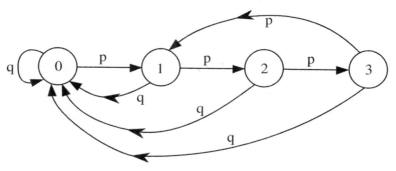

Figure 7.2: Occurrence probability Markov chain for runs of length 3.

$$\mathbf{M} = \begin{pmatrix} q & p & 0 & 0 \\ q & 0 & p & 0 \\ q & 0 & 0 & p \\ q & p & 0 & 0 \end{pmatrix} \quad (7.2)$$

Just as before, the probability of moving from s_i to s_j

in n tosses is given by $\mathbf{M}_{ij}^n$. The probability that a run occurs on toss n is equal to the probability that we are in state s_3 after toss n and since we always start in state s_0, the probability that a run occurs is simply $u_n = \mathbf{M}_{03}^n$.

There is much more that the Markov chain formalism can tell you. You can for instance get the generating functions for f_n and u_n from the matrices in equations 7.1 and 7.2 respectively. The recursion equations can also be easily written down just by looking at unique paths in the state transition diagram. The Markov matrix can give you additional probabilities that might be of interest. You can for example determine the probability of being 1 or 2 heads away from having a run and so on.

7.2   Head or Tail Run Markov Chain

Now we want to look at the slightly more complex example of recurrent runs of heads or tails. There are now two types of states that correspond to the number of consecutive heads or tails that have occurred in the toss sequence and they are labeled accordingly h_1 to h_r and t_1 to t_r. Being in state h_k means that k heads have occurred so far in the toss sequence and being in state t_k means k tails have occurred. The Markov chain in this case describes the situation after the first toss has already occurred. After the first toss we are in state h_1 with probability p or state t_1 with probability $q = 1 - p$. The first toss establishes the initial state. From state h_k we can move to h_{k+1} with probability p or to t_1 with probability q. From state t_k we can move to t_{k+1} with probability q or to h_1 with probability p. If we end up in state h_r or t_r a run has occurred and we remain there. Figure 7.3 shows the state transition diagram and the Markov matrix for head or tail runs of length 3.

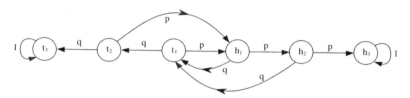

Figure 7.3: Recurrence time probability Markov chain for head or tail runs of length 3.

$$\mathbf{M} = \begin{pmatrix} 0 & q & p & 0 & 0 & 0 \\ p & 0 & 0 & q & 0 & 0 \\ 0 & q & 0 & 0 & p & 0 \\ p & 0 & 0 & 0 & 0 & q \\ 0 & 0 & 0 & 0 & 1 & 0 \\ 0 & 0 & 0 & 0 & 0 & 1 \end{pmatrix} \qquad (7.3)$$

The rows (top to bottom) and columns (left to right) correspond to the states in the following order $h_1, t_1, h_2, t_2, h_3, t_3$. The matrix element $\mathbf{M}_{13}$ for example gives the probability, q, of transitioning from t_1 to t_2 on a toss. In general the matrix element in row i and column j of the nth power of M will give the probability, after toss n, of being in the state corresponding to column j given that we started in the state corresponding to row i. We want the probability, f_n, of a head or tail run occurring for the first time on toss n. There are two ways that this can happen. At toss $n-1$, either we are in h_2 and get another head or we are in t_2 and get another tail. The probability of being in h_2 is $p\mathbf{M}_{02}^{n-2} + q\mathbf{M}_{12}^{n-2}$ and the probability of being in t_2 is $p\mathbf{M}_{03}^{n-2} + q\mathbf{M}_{13}^{n-2}$, therefore f_n is

$$\begin{aligned} f_n &= p(p\mathbf{M}_{02}^{n-2} + q\mathbf{M}_{12}^{n-2}) + q(p\mathbf{M}_{03}^{n-2} + q\mathbf{M}_{13}^{n-2}) \quad (7.4)\\ &= p^2\mathbf{M}_{02}^{n-2} + pq(\mathbf{M}_{12}^{n-2} + \mathbf{M}_{03}^{n-2}) + q^2\mathbf{M}_{13}^{n-2} \end{aligned}$$

To calculate the probability that a run of heads or tails occurs on toss n, we have to use the slightly modified chain shown in figure 7.4.

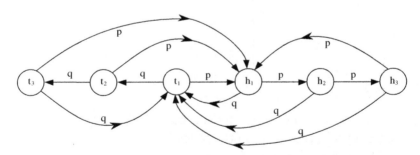

Figure 7.4: Occurrence probability Markov chain for head or tail runs of length 3.

Now the h_3 and t_3 states are no longer absorbing. You can transition from h_3 to h_1 with probability p or to t_1 with probability q and similarly for t_3. The new Markov matrix is

$$\mathbf{M} = \begin{pmatrix} 0 & q & p & 0 & 0 & 0 \\ p & 0 & 0 & q & 0 & 0 \\ 0 & q & 0 & 0 & p & 0 \\ p & 0 & 0 & 0 & 0 & q \\ p & q & 0 & 0 & 0 & 0 \\ p & q & 0 & 0 & 0 & 0 \end{pmatrix} \tag{7.5}$$

A run occurs at toss n if you are in h_3 or t_3. The probability of being in h_3 is $p\mathbf{M}_{04}^{n-1} + q\mathbf{M}_{14}^{n-1}$ and the probability of being in t_3 is $p\mathbf{M}_{05}^{n-1} + q\mathbf{M}_{15}^{n-1}$, therefore u_n is

$$u_n = p(\mathbf{M}_{04}^{n-1} + \mathbf{M}_{05}^{n-1}) + q(\mathbf{M}_{14}^{n-1} + \mathbf{M}_{15}^{n-1}) \tag{7.6}$$

7.3   Pattern Markov Chain

In the final Markov chain example we will look at recurrence times and occurrence probabilities for patterns. We are going to look at the specific pattern of $HTHTH$ but it should be obvious how you can do the same for any pattern. The set up is similar to the simple head run example discussed above. We have a state s_0 that we always start from and states s_1, s_2, s_3, s_4, s_5 that correspond to the successive stages in the build up of the pattern. When we get to stage s_5 the pattern is complete and we remain there. The Markov matrix and state transition diagram are shown below

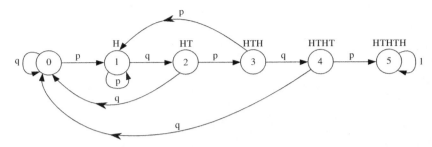

Figure 7.5: Recurrence time probability Markov chain for $HTHTH$ pattern.

$$\mathbf{M} = \begin{pmatrix} q & p & 0 & 0 & 0 & 0 \\ 0 & p & q & 0 & 0 & 0 \\ q & 0 & 0 & p & 0 & 0 \\ 0 & p & 0 & 0 & q & 0 \\ q & 0 & 0 & 0 & 0 & p \\ 0 & 0 & 0 & 0 & 0 & 1 \end{pmatrix} \qquad (7.7)$$

We are interested in the probability, f_n, that the pattern occurs for the first time on toss n. For this to happen we have to be in state s_4 at toss $n-1$ and the next toss has to be a head to complete the $HTHTH$ pattern. The probability of this happening is $f_n = p\mathbf{M}_{04}^{n-1}$.

To calculate the probability that the pattern occurs on toss n, denoted as u_n, we use the slightly modified Markov chain shown below. Now the s_5 state, where the pattern is complete, is no longer absorbing and we allow transitions out of it to s_0 if the next toss is a tail, or to s_1 if the next toss is another head. The state transition diagram and the new Markov matrix are shown below.

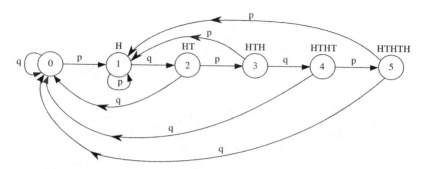

Figure 7.6: Occurrence probability Markov chain for $HTHTH$ pattern.

$$\mathbf{M} = \begin{pmatrix} q & p & 0 & 0 & 0 & 0 \\ 0 & p & q & 0 & 0 & 0 \\ q & 0 & 0 & p & 0 & 0 \\ 0 & p & 0 & 0 & q & 0 \\ q & 0 & 0 & 0 & 0 & p \\ q & p & 0 & 0 & 0 & 0 \end{pmatrix} \qquad (7.8)$$

The probability that the pattern occurs on toss n is equal to the probability that we are in state s_5 after toss n and since we always start in state s_0, the probability that a run occurs is simply $u_n = \mathbf{M}_{05}^n$.

Jamestown 400th Anniversary Commemorative Silver Dollar
Coin.
Image credit: United States Mint.

In this chapter we are going to look at probability distributions related to runs in a sequence of coin tosses. We'll start by deriving the probability generating function for the number of runs. This will give us a formula for the average and standard deviation of the number of runs. We will do the same analysis for a binary Markov chain. These results could be used to decide if a binary sequence is better described as a Markov chain or a simple coin toss. We also look at the statistics for runs of a particular length.

Then we will look at probabilities for getting a longest run of a particular length. These results could be used to infer the probability of heads and tails given the length of their longest runs. We end with a simple look at longest runs from a more combinatorial point of view.

We have by no means covered all aspects of run distributions in this chapter. That could be a book of its own. But the chapter does provide all the tools and examples necessary for any kind of run analysis you may want to do.

8.1   Number of runs in a sequence

The generating function for the number of runs in a sequence of n coin tosses can be derived from the automaton shown in figure 8.1. The automaton transitions are labeled by their weights. h represents a head, t a tail, and y a run.

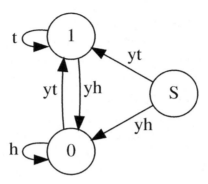

Figure 8.1: Automaton for runs generating function.

The start state is S. From there an h takes us to state 0 and a t takes us to state 1. In either case, the transition marks the begining of a run so there is a y multiplying the h and t. In state 0 another h loops back to state 0 without starting a new run so the weight of that transition is just h. If the automaton is in state 0 and outputs a t it transitions to state 1 and a new run is started so the weight of the transition is yt. The state 1 transitions are weighted similarly.

We can get a generating function that counts the number of runs in a sequence from the transition matrix for the automaton. The transition matrix is

$$\mathbf{A} = \begin{pmatrix} 0 & yh & yt \\ 0 & h & yt \\ 0 & yh & t \end{pmatrix} \tag{8.1}$$

The first row and column is for state S. The second row and column is for state 0, and the last row and column is for state 1. To get the generating function, we calculate the inverse matrix

$$\mathbf{B} = (\mathbf{I} - \mathbf{A})^{-1} \qquad (8.2)$$

Since the automaton starts in state S and can end in any of the states, the generating function is the sum of the first row of $\mathbf{B}$. This counts sequences that start at state S and end at S, 0 or 1 which allows for the empty sequence. The generating function is then:

$$g(y, h, t) = \frac{(yh - h + 1)(yt - t + 1)}{1 - h - t + ht - hty^2} \qquad (8.3)$$

If we're not interested in a specific number of heads and tails, then we can set both h and t equal to x to get the generating function $f(x, y) = g(y, x, x)$ for the number of runs in a series of n tosses.

$$f(x, y) = \frac{1 - (1 - y)x}{1 - (1 + y)x} \qquad (8.4)$$

You can readily verify that the coefficient of x^n in the expansion of $f(x, y)$ with respect to x is

$$2y(y + 1)^{n-1} \qquad (8.5)$$

This is the generating function for the number of runs in n tosses of a coin. For example, for $n = 3$ we have $2y(y + 1)^2 = 2y^3 + 4y^2 + 2y$ which indicates that there are two ways to get three runs: hth, tht. There are four ways to get two runs: htt, thh, hht, tth. And there are of course only two ways to get a single run: hhh, ttt. For $n = 4$, we have $2y(y+1)^3 = 2y^4 + 6y^3 + 6y^2 + 2y$ which indicates there are two ways to get four runs, and two ways to get a single run. These are obvious. There are six ways to get three runs, and six ways to get two runs. The six ways to get three runs are: $hthh$, $hhth$, $thtt$, $ttht$, $htth$, $thht$. In general, the generating function in equation 8.5 tells us that for n tosses, the number of ways to get $k = 1, 2, \ldots, n$ runs is

$$2\binom{n-1}{k-1} \tag{8.6}$$

This formula has a simple combinatorial interpretation. The grouping of the n tosses into k runs is equivalent to finding a compostion of the number n into k nonzero parts. For example the $n = 4$ length sequence $hthh$ corresponds to the composition $4 = 1 + 1 + 2$. It is a simple combinatorial exercise[1] to show that the number of compositions of n into k nonzero parts is $\binom{n-1}{k-1}$. The factor of 2 comes from the fact that a run sequence can start with heads or tails. For example the sequences $hthh$ and $thtt$ both correspond to the composition $4 = 1 + 1 + 2$.

8.1.1 Probability generating function

For a fair coin, the probability of getting k runs in n tosses is just equation 8.6 divided by 2^n. For the more general

---

[1]For an introduction to combinatorial problem solving see our book "Combinatorics: Problems and Solutions" listed in the references.

case of a biased coin, if p is the probability of heads, and $q = 1 - p$ is the probability of tails, then we get the probability generating function for the number of runs by letting $h = px$, and $t = qx$ in equation 8.3 for $g(y, h, t)$. Let $r(x, y, p) = g(y, px, qx)$ be the probability generating function for the number of runs, then

$$r(x, y, p) = \frac{(pxy - px + 1)(qxy - qx + 1)}{1 - x + pqx^2(1 - y^2)} \qquad (8.7)$$

Expand this equation with respect to x and let a_n be the coefficient of x^n, then

$$
\begin{aligned}
a_1 &= y \\
a_2 &= 2pqy^2 + (p^2 + q^2)y \\
a_3 &= pqy^3 + 2pqy^2 + (p^3 + q^3)y
\end{aligned}
\qquad (8.8)
$$

The a_n coefficients are probability generating functions for the number of runs in n tosses. They obey a relatively simple recurrence equation

$$a_n = a_{n-1} - pq(1 - y^2)a_{n-2} \qquad (8.9)$$

In the case of a fair coin, where $p = 1/2$, the above expressions simplify to

$$r(x, y, 1/2) = \frac{2 - (1 - y)x}{2 - (1 + y)x} \qquad (8.10)$$

The first few a_n terms for a fair coin are:

$$a_1 = y \tag{8.11}$$
$$a_2 = y(y+1)/2$$
$$a_3 = y(y+1)^2/4$$
$$a_4 = y(y+1)^3/8$$

The a_n in this case obey a simple one term recurrence

$$a_n = \frac{1}{2}(y+1)a_{n-1} \tag{8.12}$$

The solution of this recurrence is just equation 8.5 divided by 2^n

$$a_n = \frac{y(y+1)^{n-1}}{2^{n-1}} \tag{8.13}$$

8.1.2 Expected number of runs

We can get the generating function for the expected number of runs in n coin tosses by taking the derivative of $r(x,y,p)$ with respect to y and evaluating the result at $y = 1$. The coefficient of x^n in the power series expansion is then the expected number of runs in n tosses.

$$\mu(x,p) = \left. \frac{\partial}{\partial y} r(x,y,p) \right|_{y=1} \tag{8.14}$$
$$= \frac{x + (2pq-1)x^2}{(1-x)^2}$$
$$= \frac{2pq}{(1-x)^2} + \frac{1-4pq}{1-x} + 2pq - 1$$
$$= \mu_1 x + \mu_2 x^2 + \mu_3 x^3 + \cdots$$

Expanding each of the terms gives us the following formula for the expectation.

$$\mu_n = 2pq(n-1) + 1 \tag{8.15}$$

For a fixed value of n, this equation has a maximum at $p = 1/2$. You can expect the largest number of runs when using a fair coin.

The probability generating function $r(x, y, p)$ allows us to easily find the variance and standard deviation of the number of runs in n tosses. Let R_n represent the number of runs in n tosses, then the variance of R_n is

$$\sigma_n^2 = E[R_n^2] - \mu_n^2 \tag{8.16}$$

The generating function for $E[R_n^2]$ is equal to

$$\frac{\partial}{\partial y}\left(y\frac{\partial}{\partial y}r(x, y, p)\right)\bigg|_{y=1} \tag{8.17}$$

Working this out and substituting into equation 8.16 produces the following formula

$$\sigma_n^2 = 2pq(2n-3) - 4p^2q^2(3n-5) \tag{8.18}$$

A plot of the expectation and standard deviation for the number of runs in $n = 10, 20, 30, 40$ tosses is shown in figures 8.2 and 8.3. It is interesting to note that the standard deviation is not a maximum for a fair coin, and is a function of the number of tosses.

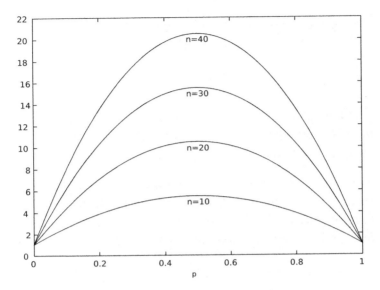

Figure 8.2: Expected number of runs.

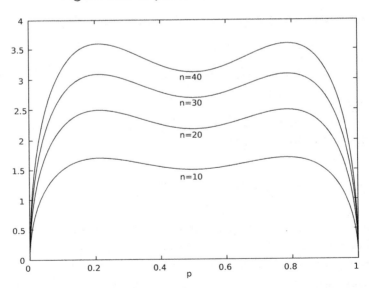

Figure 8.3: Standard deviation of number of runs.

8.1.3 Runs in a Markov Chain

Now we want to look at a more general binary Markov chain. What we have in this case is essentially two coins. One where the probability of heads is $1-a$, and the other where the probability of tails is $1-b$. When $a+b=1$ this reduces to a single coin toss. The model is represented in figure 8.4. A head puts us in state H where the probability of heads is $1-a$, and the probability of tails is a. A tail puts us in state T where the probability of tails is $1-b$ and the probability of heads is b. We can also think of a as the probability of starting a run of tails and b as the probability of starting a run of heads.

To find the run generating function for this model we use the same automaton as shown in figure 8.1, but now we have to add the probabilities to the transition weights as shown in figure 8.5.

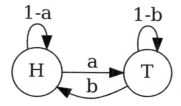

Figure 8.4: Binary Markov transition diagram.

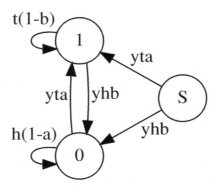

Figure 8.5: Binary Markov automaton.

The transition matrix for this automaton is

$$\mathbf{A} = \begin{pmatrix} 1 - a - b & yhb & yta \\ 0 & h(1-a) & yta \\ 0 & yhb & t(1-b) \end{pmatrix} \qquad (8.19)$$

To get the run generating function, we calculate $\mathbf{B} = (\mathbf{I} - \mathbf{A})^{-1}$ and sum the first row. Letting $h = t = x$, the generating function is

$$r(x,y,a,b) = \frac{(axy - (1-b)x + 1)(bxy - (1-a)x + 1)}{(a+b)(1 - (2-a-b)x - (aby^2 - ab + a + b - 1)x^2)}$$

$$(8.20)$$

Expand this equation with respect to x and let a_n be the coefficient of x^n, then we get the following recurrence

$$a_1 = y \tag{8.21}$$
$$a_2 = (2aby^2 + (a + b - 2ab)y)/(a + b)$$
$$a_n = (2 - a - b)a_{n-1} + (aby^2 - (1 - a)(1 - b))a_{n-2}$$

The a_n coefficients are probability generating functions for the number of runs in n iterations of the Markov chain.

8.1.4 Markov chain expected number of runs

The expected number of runs is found the same way as for the simple coin toss, i.e. take the derivative of equation 8.20 with respect to y and set $y = 1$. This gives us the following formula for the expected number of runs in n iterations of the binary Markov model

$$\mu_n = \frac{2ab}{a + b}(n - 1) + 1 \tag{8.22}$$

Comparing this with equation 8.15, we see that the two are equal when $a + b = 1$, $a = p$, $b = q$.

Contour plots of equation 8.22 for $n = 10$ and $n = 40$ are shown in figures 8.6 and 8.7. They show that the number of runs increases rapidly when a and b get close to 1, which you would expect, since this increases the probability of a switch from heads to tails or tails to heads i.e. the start of a new run.

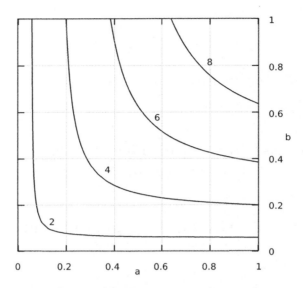

Figure 8.6: Binary Markov expected runs for $n = 10$.

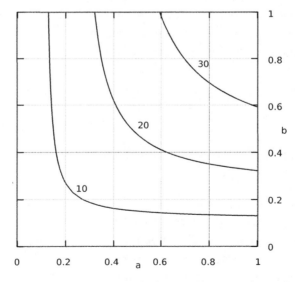

Figure 8.7: Binary Markov expected runs for $n = 40$.

The probability distributions for $n = 20$ iterations and $a = b = 1/4$, $a = b = 1/2$, $a = b = 3/4$ are plotted in figures 8.8, 8.9 and 8.10. For $a = b = 1/2$, which is equivalent to a fair coin, the expected number of runs is 10.5 and the standard deviation is 2.179449. For $a = b = 1/4$ we would expect the number of runs to be less than for a fair coin and indeed the expected number of runs is 5.75 which is almost half that for a fair coin. The standard deviation is also lower at 1.8874586. For $a = b = 3/4$ we would expect the number of runs to be greater than that for a fair coin and indeed the expected number of runs is 15.25. The standard deviation is 1.8874586, the same as for the $a = b = 1/4$ chain.

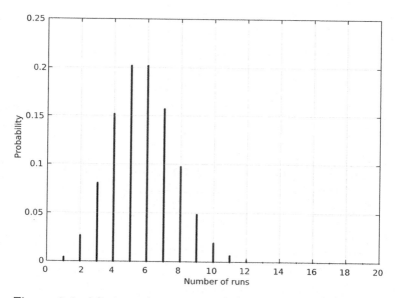

Figure 8.8: Markov chain probability distribution for number of runs for $n = 20$ and $a = b = 1/4$.

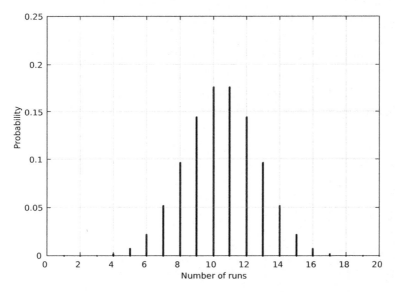

Figure 8.9: Markov chain probability distribution for number of runs for $n = 20$ and $a = b = 1/2$ (equivalent to a fair coin).

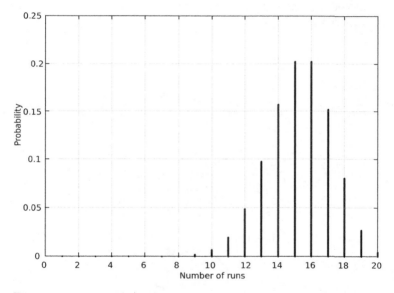

Figure 8.10: Markov chain probability distribution for number of runs for $n = 20$ and $a = b = 3/4$.

8.2 Number of runs of a given length

We've been looking at the statistics of runs in general. Now we want to look at runs of a particular length. We'll start by counting the number of binary sequences of length n that have k runs of length 1. Those sequences can be counted using a de Bruijn graph of order 2 as shown in figure 8.11.

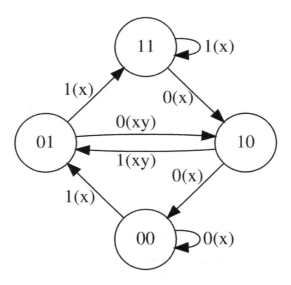

Figure 8.11: De Bruijn graph for counting sequences of runs of length 1.

Transitions are labeled by output symbol with the weight in parentheses. There is a start state (not shown) that connects to each state. All states connect to an end state (not shown) with weights given by the transition matrix

A in equation 8.23.

$$
\mathbf{A} = \begin{pmatrix}
0 & x^2 & x^2y & x^2y & x^2 & 1+2yx \\
0 & x & x & 0 & 0 & 1 \\
0 & 0 & 0 & xy & x & y \\
0 & x & xy & 0 & 0 & y \\
0 & 0 & 0 & x & x & 1 \\
0 & 0 & 0 & 0 & 0 & 0
\end{pmatrix}
\tag{8.23}
$$

The rows from top to bottom and the columns from left to right represent, in order, the states S, 00, 01, 10, 11, and E, where S and E are the start and end states. Output symbols are given a weight of x and runs are given a weight of y. Looking at the first row, 0 indicates no transition from S to itself, x^2 indicates a transition from S to 00, x^2y indicates a transition from S to 01 and the creation of a run of 0. Similarly, for the following x^2y and x^2. The $1 + 2yx$ entry in the first row indicates a connection from S to E with no output, i.e. the empty string, or the output of a single 0 or 1, both of which produce a run of length 1.

The generating function for the number of runs of length 1 can now be found by calculating $\mathbf{B} = (\mathbf{I} - \mathbf{A})^{-1}$. The element $B_{1,6}$ of this matrix is the generating function for sequences that start at S and end at E. The generating function is

$$
g_1(x, y) = \frac{1 - (1-y)x + (1-y)x^2}{1 - (1+y)x - (1-y)x^2}
\tag{8.24}
$$

The coefficient of $y^k x^n$ in the power series expansion of $g_1(x, y)$ is the number of binary strings of length n that have k runs of length 1. Those numbers are shown in table 8.1. Note that each row sums up to 2^n as it should.

		0	1	2	k 3	4	5	6	7
	0	1							
	1	0	2						
	2	2	0	2					
	3	2	4	0	2				
	4	4	4	6	0	2			
	5	6	10	6	8	0	2		
	6	10	16	18	8	10	0	2	
n	7	16	30	30	28	10	12	0	2
	8	26	52	62	48	40	12	14	0
	9	42	92	114	108	70	54	14	16
	10	68	160	216	208	170	96	70	16
	11	110	278	398	418	340	250	126	88
	12	178	480	732	808	720	516	350	160
	13	288	826	1332	1560	1450	1146	742	472
	14	466	1416	2410	2968	2922	2392	1722	1024
	15	754	2420	4332	5606	5780	5010	3710	2476
	16	1220	4124	7746	10496	11340	10260	8050	5488
	17	1974	7010	13782	19516	22010	20830	17010	12304
	18	3194	11888	24414	36048	42370	41760	35630	26752

Table 8.1: Number of binary strings of length n that have k runs of length 1.

To get the probabilities, divide the entries in each row by 2^n. Doing this, you can find the most probable number of runs of length 1 as shown in table 8.3.

The generating function for binary strings with no runs of length 1 is simply

$$g_1(x, 0) = \frac{1 - x + x^2}{1 - x - x^2} \qquad (8.25)$$

n \ k	8	9	10	11	12	13	14	15	16	17	18
8	2										
9	0	2									
10	18	0	2								
11	18	20	0	2							
12	108	20	22	0	2						
13	198	130	22	24	0	2					
14	618	240	154	24	26	0	2				
15	1368	790	286	180	26	28	0	2			
16	3438	1780	990	336	208	28	30	0	2		
17	7818	4640	2266	1220	390	238	30	32	0	2	
18	18072	10800	6116	2832	1482	448	270	32	34	0	2

Table 8.2: Continuation of Table 8.1.

For exactly one run of length 1, we have the generating function

$$\left.\frac{\partial g_1}{\partial y}\right|_{y=0} = \frac{2x(1-x)^2}{(1-x-x^2)^2} \qquad (8.26)$$

For exactly two runs of length 1, we have the generating function

$$\left.\frac{1}{2}\frac{\partial^2 g_1}{\partial y^2}\right|_{y=0} = \frac{2x^2(1-x)^3}{(1-x-x^2)^3} \qquad (8.27)$$

In general, the generating function for exactly m runs of length 1 is given by

$$\left.\frac{1}{m!}\frac{\partial^m g_1}{\partial y^m}\right|_{y=0} = \frac{2x^m(1-x)^{m+1}}{(1-x-x^2)^{m+1}} \qquad (8.28)$$

The same basic procedure can be used to find the generating function that counts binary sequences with runs of

136

n	Number of Runs	Probability
0	0	1.0
1	1	1.0
2	0,2	0.5
3	1	0.5
4	2	0.375
5	1	0.3125
6	2	0.28125
7	1,2	0.23438
8	2	0.24219
9	2	0.22266
10	2	0.21094
11	3	0.2041
12	3	0.19727
13	3	0.19043
14	3	0.18115
15	4	0.17639
16	4	0.17303
17	4	0.16792
18	4	0.16163
19	5	0.15821
20	5	0.15569

Table 8.3: Most probable number of runs of length 1 when tossing a fair coin n times.

length l.

$$g_l(x,y) = \frac{1 - (1-y)x^l + (1-y)x^{l+1}}{1 - 2x + (1-y)x^l - (1-y)x^{l+1}} \qquad (8.29)$$

The coefficient of $y^k x^n$ in the power series expansion of $g_l(x,y)$ is the number of binary sequences of length n that have k runs of length l.

The generating function for exactly $m > 0$ runs of length l is shown in equation 8.30.

$$\frac{1}{m!} \frac{\partial^m g_l}{\partial y^m}\bigg|_{y=0} = \frac{2x^{ml}(1-x)^{m+1}}{(1 - 2x + x^l - x^{l+1})^{m+1}} \qquad (8.30)$$

The following identity is easy to verify.

$$\sum_{m=0}^{\infty} \frac{1}{m!} \frac{\partial^m g_l}{\partial y^m}\bigg|_{y=0} = \frac{1}{1 - 2x} \qquad (8.31)$$

The expression on the right of this equation is the generating function for the number of binary strings of length n. The equation says that every binary sequence must have some number of runs of length l (that number may of course be 0).

From $g_l(x,y)$ we can get the generating function for the expected number of runs of length l in n tosses of a fair coin. The generating function is shown in equation 8.32.

$$\frac{\partial g_l(x/2, y)}{\partial y}\bigg|_{y=1} = \frac{(x-2)^2 x^l}{(x-1)^2} \tag{8.32}$$

From equation 8.32 we get the following expression for the expectation

$$\mu_{n,l} = \begin{cases} 2/2^l & n = l \\ (n - l + 3)/2^{l+1} & n > l \end{cases} \tag{8.33}$$

Table 8.4 shows these expectations for $n = 10$ through 40 and $l = 1$ through 5.

8.3 Longest Runs

Now we turn our attention to the longest run of heads and tails in a binary sequence. What for example is the probability that when a coin is tossed 50 times the longest run of heads is 6. What is the probability that both the longest run of heads and tails is 6? We will show how to answer these kinds of questions.

There are several ways to approach this type of problem. The easiest way is to start with the run automaton shown in figure 8.12.

There are two outputs for the automaton and each output represents a run. The output a represents a run of 0's,

		l				
		1	2	3	4	5
	10	3.0	1.375	0.625	0.28125	0.125
	11	3.25	1.5	0.6875	0.3125	0.140625
	12	3.5	1.625	0.75	0.34375	0.15625
	13	3.75	1.75	0.8125	0.375	0.171875
	14	4.0	1.875	0.875	0.40625	0.1875
	15	4.25	2.0	0.9375	0.4375	0.203125
	16	4.5	2.125	1.0	0.46875	0.21875
	17	4.75	2.25	1.0625	0.5	0.234375
	18	5.0	2.375	1.125	0.53125	0.25
	19	5.25	2.5	1.1875	0.5625	0.265625
	20	5.5	2.625	1.25	0.59375	0.28125
	21	5.75	2.75	1.3125	0.625	0.296875
	22	6.0	2.875	1.375	0.65625	0.3125
n	23	6.25	3.0	1.4375	0.6875	0.328125
	24	6.5	3.125	1.5	0.71875	0.34375
	25	6.75	3.25	1.5625	0.75	0.359375
	26	7.0	3.375	1.625	0.78125	0.375
	27	7.25	3.5	1.6875	0.8125	0.390625
	28	7.5	3.625	1.75	0.84375	0.40625
	29	7.75	3.75	1.8125	0.875	0.421875
	30	8.0	3.875	1.875	0.90625	0.4375
	31	8.25	4.0	1.9375	0.9375	0.453125
	32	8.5	4.125	2.0	0.96875	0.46875
	33	8.75	4.25	2.0625	1.0	0.484375
	34	9.0	4.375	2.125	1.03125	0.5
	35	9.25	4.5	2.1875	1.0625	0.515625
	36	9.5	4.625	2.25	1.09375	0.53125
	37	9.75	4.75	2.3125	1.125	0.546875
	38	10.0	4.875	2.375	1.15625	0.5625
	39	10.25	5.0	2.4375	1.1875	0.578125
	40	10.5	5.125	2.5	1.21875	0.59375

Table 8.4: Expected number of runs of length l when tossing a fair coin n times.

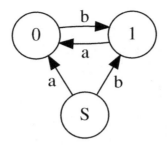

Figure 8.12: Binary run automaton.

and b represents a run of 1's.[2] The automaton's transition matrix is:

$$\mathbf{A} = \begin{pmatrix} 0 & a & b \\ 0 & 0 & b \\ 0 & a & 0 \end{pmatrix} \qquad (8.34)$$

In top to bottom and left to right order, the rows and columns represent the states S, 0, and 1. A binary sequence will start in state S and end in any of the states[3] so to get the run generating function we calculate $\mathbf{B} = (\mathbf{I} - \mathbf{A})^{-1}$ and sum the first row of $\mathbf{B}$. The generating function is then

$$g(a,b) = \frac{(1+a)(1+b)}{1-ab} \qquad (8.35)$$

The first few terms in the power series expansion of $g(a,b)$ are $1 + a + b + 2ba + ba^2 + b^2a + 2b^2a^2 + b^2a^3 + b^3a^2 + 2b^3a^3 + b^3a^4 + b^4a^3 + 2b^4a^4 + b^4a^5 + b^5a^4 + 2b^5a^5 + b^5a^6 + b^6a^5 + 2b^6a^6 + \cdots$. The $2b^3a^3$ term for example, says

---

[2]We'll be using 0's and 1's in place of heads and tails in the following discussion.

[3]We are allowing for sequences of length zero.

there are two sequences with three runs of 0 and three runs of 1, they are $ababab$ and $bababa$. The b^3a^4 term says there is only one sequence with four runs of 0 and three runs of 1, it is $abababa$.

Any binary sequence will start with either $abab\cdots$ or $baba\cdots$, so there are only two ways to have a binary sequence with n runs. We can see this from the generating function by setting $a = b = z$ to get

$$g(z, z) = \frac{1 + z}{1 - z} \tag{8.36}$$
$$= 1 + 2z + 2z^2 + 2z^3 + \cdots$$

From $g(a, b)$ we can easily get the generating function for sequences that have no runs of 0 greater than k, and no runs of 1 greater than l. First let x represent 0 and y represent 1 then in equation 8.35 let

$$a = x + x^2 + \cdots + x^k \tag{8.37}$$
$$= \frac{x(1 - x)^k}{1 - x}$$

$$b = y + y^2 + \cdots + y^l \tag{8.38}$$
$$= \frac{y(1 - y)^l}{1 - y}$$

This will limit runs of 0 to be no greater than k, and runs of 1 to be no greater than l. The generating function for the number of such sequences is then:

$$f_{k,l}(x, y) = g\left(\frac{x(1 - x^k)}{1 - x}, \frac{y(1 - y^l)}{1 - y}\right) \qquad (8.39)$$

$$= \frac{\left(x^{k+1} - 1\right)\left(y^{l+1} - 1\right)}{1 - x - y + xy^{l+1} + x^{k+1}y - x^{k+1}y^{l+1}}$$

If we don't care about the exact number of 0's and 1's we can let $x = y = z$ to get the generating function

$$f_{k,l}(z) = \frac{\left(z^{k+1} - 1\right)\left(z^{l+1} - 1\right)}{1 - 2z + z^{l+2} + z^{k+2} - z^{l+k+2}} \qquad (8.40)$$

With no restrictions on the length of 0 or 1 runs, i.e. $k = l = \infty$ we get the generating function:

$$f_{\infty,\infty}(x, y) = g\left(\frac{x}{1 - x}, \frac{y}{1 - y}\right) \qquad (8.41)$$

$$= \frac{1}{1 - x - y}$$

As you would expect, this is the generating function for the number of general binary sequences. To see this, let $x = y = z$ and you get:

$$f_{\infty,\infty}(z) = \frac{1}{1 - 2z} \qquad (8.42)$$

The coefficient of z^n in the power series expansion is 2^n, i.e. the number of binary sequences of length n.

As another example, the generating function for the number of strings with no runs of 0 greater than one and no restriction on the length of runs of 1 is:

$$f_{1,\infty}(z) = g\left(z, \frac{z}{1-z}\right) \qquad (8.43)$$

$$= \frac{1+z}{1-z-z^2}$$

This is the generating function for the well known Fibonacci numbers. The first few terms in the power series expansion are $1 + 2z + 3z^2 + 5z^3 + 8z^4 + 13z^5 + 21z^6 + 34z^7 + 55z^8 + 89z^9 + 144z^{10} + 233z^{11} + 377z^{12} + 610z^{13} + 987z^{14} + 1597z^{15} + 2584z^{16} + 4181z^{17} + 6765z^{18} + 10946z^{19} + 17711z^{20} + \cdots$. From the expansion you can see for example that out of the $2^{10} = 1024$ binary sequences of length 10 there are only 144 that have no run of 0 longer than one.

We can use the $f_{k,\infty}(z)$ functions to get the generating function for the number of sequences where the length of the longest run of 0 is exactly k and there is no restriction on the length of runs of 1. The generating function is

$$F_k(z) = f_{k,\infty}(z) - f_{k-1,\infty}(z) \qquad (8.44)$$

$$= \frac{z^k(z-1)^2}{(z^{k+1} - 2z + 1)(z^{k+2} - 2z + 1)}$$

The functions for $k = 0, 1, 2, 3$ are:

$$F_0(z) = \frac{1}{1 - z}$$

$$F_1(z) = \frac{z}{(1 - z)(1 - z - z^2)}$$

$$F_2(z) = \frac{z^2}{(1 - z - z^2)(1 - z - z^2 - z^3)}$$

$$F_3(z) = \frac{z^3}{(1 - z - z^2 - z^3)(1 - z - z^2 - z^3 - z^4)}$$

Table 8.5 shows the number of binary sequences of length n where the longest run of 0 is exactly k for $k = 0, 1, \ldots, 7$.

		0	1	2	3	4	5	6	7
	0	1							
	1	1	1						
	2	1	2	1					
	3	1	4	2	1				
	4	1	7	5	2	1			
	5	1	12	11	5	2	1		
	6	1	20	23	12	5	2	1	
	7	1	33	47	27	12	5	2	1
n	8	1	54	94	59	28	12	5	2
	9	1	88	185	127	63	28	12	5
	10	1	143	360	269	139	64	28	12
	11	1	232	694	563	303	143	64	28
	12	1	376	1328	1167	653	315	144	64
	13	1	609	2526	2400	1394	687	319	144
	14	1	986	4781	4903	2953	1485	699	320
	15	1	1596	9012	9960	6215	3186	1519	703
	16	1	2583	16929	20135	13008	6792	3277	1531
	17	1	4180	31709	40534	27095	14401	7026	3311
	18	1	6764	59247	81300	56201	30391	14984	7117
	19	1	10945	110469	162538	116143	63872	31808	15218
	20	1	17710	205606	324020	239231	133751	67249	32392

Table 8.5: Number of sequences of length n where longest run of 0 is k

You can get the probability of getting such a sequence with a fair coin by dividing each row of table 8.5 by 2^n.

You can also get the probabilities directly from the probability generating function which is just equation 8.44 evaluated at $z = z/2$. We will look at the more general case of a biased coin below.

First we should look at the generating function for the number of sequences where the length of the longest run of 0 is exactly k and the length of the longest run of 1 is exactly l. That generating function is given by

$$F_{k,l}(z) = f_{k,l}(z) - f_{k-1,l}(z) - f_{k,l-1}(z) + f_{k-1,l-1}(z)$$
(8.45)

Note that we have to add back $f_{k-1,l-1}(z)$ because those strings have been subtracted twice in the $f_{k-1,l}(z)$ and $f_{k,l-1}(z)$ subtractions. Table 8.6 shows the number of sequences of length 10 where the longest run of 0 is exactly k and the longest run of 1 is exactly l for $k,l = 1, 2, 3, 4, 5, 7, 8, 9$.

		1	2	3	4	5	6	7	8	9
	1	2	26	41	32	20	11	6	3	2
	2	26	124	109	58	26	11	4	2	0
k	3	41	109	70	31	12	4	2	0	0
	4	32	58	31	12	4	2	0	0	0
	5	20	26	12	4	2	0	0	0	0
	6	11	11	4	2	0	0	0	0	0
	7	6	4	2	0	0	0	0	0	0
	8	3	2	0	0	0	0	0	0	0
	9	2	0	0	0	0	0	0	0	0

Table 8.6: Number of sequences of length 10 where longest run of 0 is k and the longest run of 1 is l

To get the probability of getting such a sequence with a fair coin you divide the numbers in the table by $2^{10} =$

1024. Note that $k = l = 2$ is the most probable sequence with probability $124/1024 = 0.121$. The $k = 3$, $l = 2$ and $k = 2$, $l = 3$ sequences are almost as likely with probability $109/1024 = 0.106$. The probabilities of getting other values of k and l drop off rather quickly.

Now we look at the more general case of a biased coin where the probability of 0 is p and the probability of 1 is $q = 1 - p$, the probability generating function for sequences with no run of 0 greater than k and no restriction on the length of 1 runs is found from equation 8.35 as follows:

$$h_{k,\infty}(z) = g\left(\frac{pz(1 - (pz)^k)}{1 - pz}, \frac{qz}{1 - qz}\right) \qquad (8.46)$$
$$= \frac{1 - (pz)^{k+1}}{1 - z + qz(pz)^{k+1}}$$

We'll look at a couple of examples of how to use these functions. First let's find the probability that the length of the longest run of 0 in a binary sequence is exactly one. For that we need the functions $h_{0,\infty}(z)$ and $h_{1,\infty}(z)$ shown in equations 8.47 and 8.48.

$$h_{0,\infty}(z) = \frac{1 - pz}{1 - z + pqz^2} \qquad (8.47)$$
$$= \frac{1}{1 - qz}$$

$$h_{1,\infty}(z) = \frac{1 - (pz)^2}{1 - z + p^2qz^3} \qquad (8.48)$$
$$= \frac{1 + pz}{1 - qz - pqz^2}$$

The probability generating function we want is then, $H_1(z) = h_{1,\infty}(z) - h_{0,\infty}(z)$. Expanding this in a power series gives us the probabilities for sequences where the length of the longest run of 0 is exactly one. Those probabilities for $n = 1, 2, \ldots, 8$ are shown in table 8.7.

n	Probability
1	p
2	$2p - 2p^2$
3	$2p^3 - 5p^2 + 3p$
4	$-p^4 + 6p^3 - 9p^2 + 4p$
5	$-4p^4 + 13p^3 - 14p^2 + 5p$
6	$2p^5 - 12p^4 + 24p^3 - 20p^2 + 6p$
7	$p^7 - 4p^6 + 12p^5 - 29p^4 + 40p^3 - 27p^2 + 7p$
8	$-2p^8 + 10p^7 - 23p^6 + 40p^5 - 60p^4 + 62p^3 - 35p^2 + 8p$
9	$2p^9 - 15p^8 + 45p^7 - 78p^6 + 101p^5 - 111p^4 + 91p^3 - 44p^2 + 9p$
10	$-p^{10} + 14p^9 - 63p^8 + 144p^7 - 205p^6 + 216p^5 - 189p^4 + 128p^3 - 54p^2 + 10p$

Table 8.7: Probability that the longest run of 0 in a binary sequence of length n is exactly 1.

Let P_n be the probability that longest run of 0 in a binary sequence of length n is exactly 1 then the recurrence in equation 8.49 can be used to generate the probabilities for all values of n given the initial conditions P_1, P_2, and P_3 in table 8.7. The recurrence can easily be found from the definition of $H_1(z)$.

$$P_n = (1 - p)(2P_{n-1} - (1 - 2p) * P_{n-2} - p(1 - p)P_{n-3}$$
$$(8.49)$$

Plots of the probabilities in table 8.7 as a function of p where $0 \le p \le 1$ are shown in figure 8.13 for $n = 2, 3, 4, 5, 6$. Note that as n increases the peaks in these plots shift to lower values of p, i.e. the most likely value of p decreases. This agrees with the intuitive notion that to get a long binary sequence where the longest run of 0

is 1, the probability of getting a 0 must be small.

What's not so obvious is that the height of the peaks increases and narrows as n increases. If you toss a coin 10 times and you get a maximum run of 0's of length 1 then you can be fairly sure that the probability of 0 is close to the peak of the equation in row 10 of table 8.7. The certainty is not so great if you only toss the coin 3 times and get only one 0 in a row.

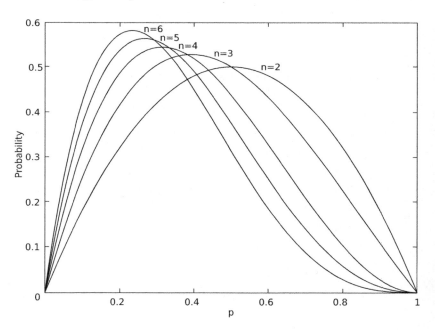

Figure 8.13: Probability that the longest run of 0 in a binary sequence of length n is exactly 1.

Now let's find the probability that the longest run of 0 in a binary sequence is exactly 2. For that we need the functions $h_{1,\infty}(z)$ and $h_{2,\infty}(z)$ shown in equations 8.48

and 8.50 respectively.

$$h_{2,\infty}(z) = \frac{1 + pz + p^2 z^2}{1 - qz - pqz^2 - p^2 qz^3} \tag{8.50}$$

The probability generating function we want is then $H_2(z) = h_{2,\infty}(z) - h_{1,\infty}(z)$. Expanding this in a power series gives us the probabilities. Those probabilities for $n = 2, 3, \ldots, 8$ are shown in table 8.8.

The probabilities can also be calculated with a recurrence relation. In this case it is easier to use a recurrence to calculate the probabilities associated with both $h_{2,\infty}(z)$ and $h_{1,\infty}(z)$ and then subtract the two. Let a_n be the $h_{2,\infty}(z)$ probabilities and b_n be the $h_{1,\infty}(z)$ probabilities, then with the initial conditions $a_0 = a_1 = a_2 = 1$ and $b_0 = b_1 = 1$, their recurrence relations are:

$$a_n = q(a_{n-1} + pa_{n-2} + p^2 a_{n-3}) \tag{8.51}$$

$$b_n = q(b_{n-1} + pb_{n-2}) \tag{8.52}$$

The probabilities in table 8.8 are then given by $P_n = a_n - b_n$. The recurrence relations for a_n and b_n come directly from the $h_{2,\infty}(z)$ and $h_{1,\infty}(z)$ generating functions.[4]

Plots of the probabilities for $n = 4, 5, 6, 7, 8$ and $0 \le p \le 1$ are shown in figure 8.14. Once again, as n increases the most likely value of p decreases but this time not so quickly. The peaks also do not narrow as quickly.

---

[4]To learn how to get recurrence relations from generating functions see the book "Generatingfunctionology" by Herbert S. Wilf listed in the references.

n	Probability
2	p^2
3	$2p^2 - 2p^3$
4	$p^4 - 4p^3 + 3p^2$
5	$p^5 + p^4 - 6p^3 + 4p^2$
6	$-p^6 + 4p^5 - 8p^3 + 5p^2$
7	$-p^7 - 2p^6 + 9p^5 - 2p^4 - 10p^3 + 6p^2$
8	$2p^8 - 6p^7 - 2p^6 + 16p^5 - 5p^4 - 12p^3 + 7p^2$
9	$-p^9 + 9p^8 - 18p^7 + 25p^5 - 9p^4 - 14p^3 + 8p^2$
10	$-4p^9 + 24p^8 - 40p^7 + 5p^6 + 36p^5 - 14p^4 - 16p^3 + 9p^2$

Table 8.8: Probability that the longest run of 0 in a binary sequence of length n is exactly 2

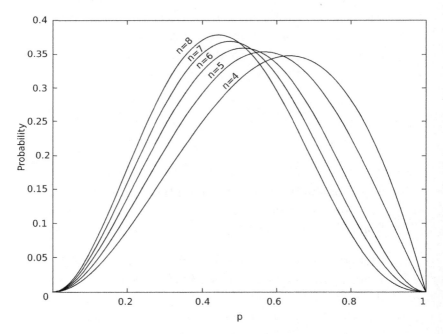

Figure 8.14: Probability that the longest run of 0 in a binary sequence of length n is exactly 2.

In general, the probability generating function for binary sequences where the longest run of 0 has length k is

$$H_k(z) = h_{k,\infty}(z) - h_{k-1,\infty}(z) \qquad (8.53)$$

The probabilities can be calculated using a recurrence for the $h_{k,\infty}(z)$ and $h_{k-1,\infty}(z)$ probabilities. For the $h_{k,\infty}(z)$ probabilities the recurrence is given by $a_0 = a_1 = \cdots = a_k = 1$ and

$$a_n = (1-p)\sum_{i=0}^{k} p^i a_{n-i-1} \qquad (8.54)$$

For the $h_{k-1,\infty}(z)$ probabilities the recurrence is given by $b_0 = b_1 = \cdots = b_{k-1} = 1$ and

$$b_n = (1-p)\sum_{i=0}^{k-1} p^i b_{n-i-1} \qquad (8.55)$$

The probability that in sequence of length n the longest run of 0 is equal to k is then $P_n = a_n - b_n$.

8.3.1 Expected Length of Longest Run

We can get the generating function for the expected length of the longest run of 0 from the $H_k(z)$ functions defined in equation 8.53. Let $E(z)$ be the generating function then

$$E(z) = \sum_{k=1}^{\infty} k H_k(z)$$

$$= \sum_{k=1}^{\infty} k(h_{k,\infty}(z) - h_{k-1,\infty}(z)) \qquad (8.56)$$

The sum in this equation cannot be easily calculated so to get it into a better form we rewrite it as follows:

$$E(z) = \lim_{n \to \infty} \left(n h_{n,\infty}(z) - \sum_{k=0}^{n-1} h_{k,\infty}(z) \right)$$

$$= \lim_{n \to \infty} \sum_{k=0}^{n-1} \left(h_{n,\infty}(z) - h_{k,\infty}(z) \right) \qquad (8.57)$$

We can take the limit in this last expression by noting that

$$\lim_{n \to \infty} h_{n,\infty}(z) = \frac{1}{1-z} \qquad (8.58)$$

The proof of this limit is as follows: $h_{n,\infty}(z)$ is the probability generating function for sequences that have no run of 0 longer than n but for sequences of length $k \leq n$ there can be no run longer than n so the probability is 1. The power series for $h_{n,\infty}(z)$ must therefor agree with the power series for $1/(1-z)$ for all the first $n+1$ terms up to z^n, i.e. the coefficients of z^k for $k = 0, 1, \ldots, n$ are equal to 1 for both power series. As $n \to \infty$ we have the equality in equation 8.58. So we can express $E(z)$ as follows

$$E(z) = \sum_{k=0}^{\infty} \left(\frac{1}{1-z} - h_{k,\infty}(z) \right) \qquad (8.59)$$

We can check this equation by looking at the cases where $p = 0$ and $p = 1$. For $p = 0$ there will be no 0's in the sequence so the expected longest run length should be 0. Indeed when $p = 0$ we have $h_{k,\infty}(z) = 1/(1-z)$ and $E(z) = 0$. For $p = 1$ there will be nothing but 0's in the sequence so the expected longest run length should be n. For $p = 1$ we have $h_{k,\infty}(z) = (1 - z^{k+1})/(1 - z)$ which when substituted into equation 8.59 results in, after simplification, $E(z) = z/(1 - z)^2$. The coefficient of z^n in the power series expansion of this expression is n.

Equation 8.59 is the simplest form in which to evaluate $E(z)$. To get the expected run length for a sequence of length n the sum only needs to be evaluated up to $k = n$. The expectation is then the coefficient of z^n in the resulting power series. The expectations for $n = 10, 20, 30, 40$ as a function of p are shown in figure 8.15

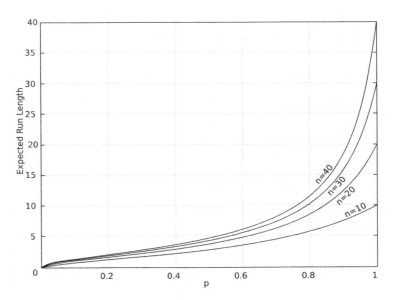

Figure 8.15: Expected length of the longest run of 0 in binary sequences of length $10, 20, 30, 40$ as a function of p.

8.4   Combinatorics of Longest Runs

In the previous section we looked a longest runs using a generating function approach which is very general and powerful and can be extended to sequences with more than two outcomes. In this section we're going to take a quick look at longest runs from a more combinatorial perspective. While not as general as the generating function approach, it is less abstract which can be useful for developing a picture of what's going on.[5]

8.4.1   Fair Coin Head Runs

For a fair coin with head and tail probabilities equal to $1/2$, We want the probability that in a sequence of n tosses, a head never comes up more than m times in a row. Let R_n be the random variable equal to the length of the longest run of heads in n tosses. R_n can range from 0 for no heads up to n for all heads. We want to find the probability that R_n is less than or equal to m. This is the cumulative distribution function for R_n which we denote as follows:

$$F_n(m) = P(R_n \le m) \qquad (8.60)$$

To calculate this we need to find how many sequences of n coin tosses have no more than m consecutive heads.

---

[5]The development is essentially the same as that in the paper by Mark F. Schilling listed in the references.

Let $A(n, m)$ be the number of such sequences then, since all sequences have a probability of $1/2^n$, we have:

$$F_n(m) = \frac{A(n, m)}{2^n} \qquad (8.61)$$

To determine $A(n, m)$ first note that if $n \leq m$ then none of the sequences can have a run of more than m heads. In this case $A(n, m)$ is just equal to the total number of sequences.

$$A(n, m) = 2^n \qquad n \leq m \qquad (8.62)$$

For n greater than m it will help to look at a specific example. In the case of $n > m = 1$ we want the number of sequences that have no more than 1 head in a row. Each of these sequences must begin with either T (a tail) or HT (a head followed by a tail). If the sequence begins with a T, remove it and you are left with a sequence of length $n - 1$ that has $m = 1$. If the sequence begins with a HT, remove it and you are left with a sequence of length $n - 2$ that has $m = 1$. Since these are the only two possibilities, it must be true that

$$A(n, 1) = A(n - 1, 1) + A(n - 2, 1) \qquad (8.63)$$

This simple recurrence can be solved by starting with two initial values from equation 8.62, $A(0, 1) = 1$ and $A(1, 1) = 2$. With equation 8.63, you then get $A(2, 1) = A(1, 1) + A(0, 1) = 2 + 1 = 3$ and so on for larger values of n. The following table shows the values up to $n = 10$. This happens to be the Fibonacci sequence.

n	0	1	2	3	4	5	6	7	8	9	10
$A(n, 1)$	1	2	3	5	8	13	21	34	55	89	144

Table 8.9: Evaluating recurrence equation 8.63 with initial values $A(0, 1) = 1$ and $A(1, 1) = 2$ for n up to 10.

The same argument applies to larger values of m. For $m = 2$ we want the number of sequences with no more than 2 heads in a row. Each sequence will begin with either T, HT, or HHT. Deleting these we have sequences of length $n - 1$, $n - 2$, and $n - 3$ all of which have no more than 2 heads in a row. Therefore we must have

$$A(n, 2) = A(n-1, 2) + A(n-2, 2) + A(n-3, 2) \quad \text{(8.64)}$$

Generalizing this to any value of m gives the following recurrence equation:

$$A(n, m) = \sum_{i=1}^{m+1} A(n - i, m) \qquad n > m \qquad \text{(8.65)}$$

where the initial $m+1$ values are calculated using equation 8.62.

8.4.2 Fair Coin Head or Tail Runs

Now let's look at the probability that the longest run of heads or tails in a sequence of n tosses is less than or equal to m. This can be turned into an equivalent problem involving just head runs. To see how, look at the example of 40 coin tosses shown below.

TTTTHHHTHHTTTTHTHTHHTTTTHTHTTHHHHHHTTTTTH
HHHTHHTTHTHHTTTTTHTHHHTTTTHTHHHHHHTHHHHT

The first row shows the tosses. The second row encodes the runs of both heads and tails by replacing each pair of tosses with a H if they are equal and a T if they are different. A run of H or T that has length k becomes a run of H with length $k - 1$ in the second row. Let $B(n, m)$ be the number of sequences of n tosses that have no runs of heads or tails longer than m. From the encoding example above it would seem that $B(n, m)$ is equal to the number of sequences of $n - 1$ tosses that have no head runs longer than $m - 1$. It is actually twice this number since if we exchange heads and tails in the toss sequence we get the same run encoding sequence. We therefore have:

$$B(n, m) = 2A(n - 1, m - 1) \qquad (8.66)$$

Let the random variable R_n now represent the longest run of heads or tails. The probability distribution function for R_n is then:

$$F_n(m) = P(R_n \leq m) \qquad (8.67)$$
$$= \frac{B(n, m)}{2^n}$$
$$= \frac{2A(n-1, m-1)}{2^n}$$
$$= \frac{A(n-1, m-1)}{2^{n-1}}$$

8.4.3 Biased Coin Head Runs

For a biased coin all sequences no longer have the same probability. The probability depends on exactly how many heads and tails occur. The sequences of n tosses in which no more than m heads appear in a row now have to be split into sequences containing $k = 0, 1, 2, \ldots, n$ heads. In the trivial case $n \leq m$ so that all sequences meet the criteria of having head runs less than or equal to m. The number of sequences with k heads is $\binom{n}{k}$ and we have

$$F_n(m) = P(R_n \leq m) \qquad (8.68)$$
$$= \sum_{k=0}^{n} \binom{n}{k} p^k q^{n-k}$$
$$= (p + q)^n$$
$$= 1$$

Where p is the probability of heads and $q = 1 - p$ is the probability of tails. In other cases let $C(n, k, m)$ be the number of sequences with exactly k heads and no more than m of them in a row. The total number of sequences with no run of more than m heads is then

$$A(n,m) = \sum_{k=0}^{n} C(n,k,m) \qquad (8.69)$$

and the cumulative distribution function is

$$F_n(m) = \sum_{k=0}^{n} C(n,k,m) p^k q^{n-k} \qquad (8.70)$$

The $C(n,k,m)$ can be calculated recursively in a manner similar to the way the $A(n,m)$ are calculated. First note that for $k \le m$ there can be no run greater than m so that $C(n,k,m) = \binom{n}{k}$ and for $n > m$ we have $C(n,n,m) = 0$. We therefore have to calculate $C(n,k,m)$ for $m < k < n$. For $m = 1$ each contributing sequence begins with T or HT therefore

$$C(n,k,1) = C(n-1,k,1) + C(n-2,k-1,1) \quad (8.71)$$

For $m = 2$ the contributing sequences begin with T, HT, or HHT so that

$$C(n,k,2) = \qquad\qquad\qquad\qquad\qquad (8.72)$$
$$C(n-1,k,2) + C(n-2,k-1,2) + C(n-3,k-2,2)$$

In general then we have

$$C(n, k, m) = \sum_{i=0}^{m} C(n - 1 - i, k - i, m) \qquad (8.73)$$

Problem 1. If you toss a coin 30 times, why are you more likely to get 15 heads and 15 tails, than 30 heads?

Answer. There is only one way to get 30 heads in 30 coin tosses. There are $\binom{30}{15} = 155,117,520$ ways to get 15 heads and 15 tails. There are a total of $2^{30} = 1,073,741,824$ different possible results of 30 coin tosses, so the probability of getting 15 heads and 15 tails is $\binom{30}{15}/2^{30} = 0.144$ while the probability of getting 30 heads is $1/2^{30} = 0.000000000931$, which makes it 155 million times more likely to get 15 heads than 30 heads.

Problem 2. If you toss a fair coin an even number of times, what is the probability of getting an equal number of heads and tails?

Answer. When you toss a coin $2n$ times, there are $\binom{2n}{n}$ ways to get n heads and n tails. There are 2^{2n} total ways to toss a coin $2n$ times, so the probability of getting n heads and n tails in $2n$ tosses of a fair coin is $\binom{2n}{n}/2^{2n}$. This probability can also be gotten from the binomial distribution equation (equation 2.5), for $P(2n, n)$ with $p = 1/2$. The probabilities for $n = 1, 2 \ldots 8$ are shown in table 9.1. Figure 9.1 shows the probability of an equal number of heads and tails versus the probability of heads for $n = 2, 4, 6, \ldots 20$ tosses. Note that as the number of tosses increases, the probability of getting an equal

161

n	1	2	3	4	5	6	7	8
prob	0.5	0.375	0.313	0.273	0.246	0.226	0.209	0.196

Table 9.1: Probability of getting n heads and n tails in $2n$ tosses of a fair coin.

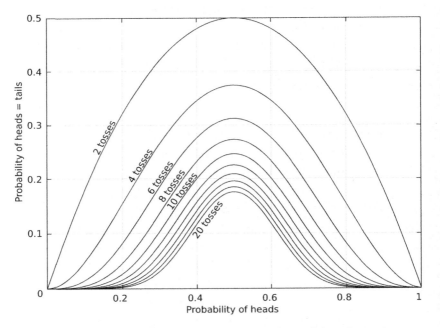

Figure 9.1: Probability of an equal number of heads and tails versus the probability of heads for $n = 2, 4, 6, \ldots 20$ tosses.

number of heads and tails goes down, and the curve narrows.

Problem 3. Is it unusual to get only 6 heads in 20 tosses of a fair coin?

Answer. From equation 3.8, the expectation is $E = np = 20 \cdot 0.5 = 10$, and the variance is at most $Var = np(1 - p) = 20 \cdot 0.5 \cdot 0.5 = 5$. Since 6 is within ± 5 of the expectation 10, it is not unusual to get only 6 heads in 20 tosses of a fair coin.

Problem 4. If you have a coin that you are uncertain is fair, how can you best estimate its probability of heads?

Answer. Assuming you have a record of tossing a coin n times, and heads showed up k times where $0 < k < n$, according to Laplace's rule of succession, you should estimate the probability of heads by using the mean which is $(k+1)/(n+2)$. If more than a point estimate is needed, the probability distribution in this case is given by equation 3.19, and is $(n + 1)\binom{n}{k}p^k(1 - p)^{n-k}$ where p is the probability of heads.

Problem 5. In 100 tosses of a fair coin, what is the probability there will be no more than 40 heads?

Answer. An exact calculation can be done using equation 3.6 with $n = 100$, $m = 40$, and $p = 0.5$, giving $\sum_{k=0}^{40} \binom{100}{k}/2^{100} = 0.028444$. For the normal approximation, we can use the QuantWolf Normal Distribution Calculator [1] with $\mu = 50$, $\sigma = 5$, and $a = m + 1/2 = 40.5$ to get 0.028716. We can get both the exact and normal approximation probabilities using the QuantWolf Binomial Distribution Calculator [2] with $p = 0.5$, $n = 100$, and $k = 40$.

Problem 6. What is the probability that a fair coin has to be tossed 5 times before the first head appears?

Answer. We can use equation 3.29 with $p = 0.5$ and the geometric random variable X with value $m = 5$ giving us $P(X = 5) = 0.5 \cdot (1 - 0.5)^4 = 0.03125$. Note that this probability can be expressed as $1/32$, so we can say that the odds are 31 to 1 against it happening.

Problem 7. What is the probability that it takes 10 tosses to get 3 heads with a fair coin?

Answer. We can answer this with the formula for the negative binomial distribution, equation 3.33, with $p = 0.5$, $r = 3$ and the negative binomial random variable X with value $m = 10$ giving us $P(X = 10) = \binom{9}{2} 0.5^3 \cdot (1 - 0.5)^7 = 0.035156$.

Problem 8. How is the coin toss related to the amount of time it takes for an unstable nucleus to decay?

---

[1] http://www.quantwolf.com/calculators/normalprobcalc.html
[2] http://www.quantwolf.com/calculators/binomialdistcalc.html

Answer. The continuous limit of the expression for the probability of getting the first heads sometime in the first n tosses also gives the probability that an unstable nucleus decays within a given time. That expression is given by equation 3.41 and is $F(t) = 1 - e^{-\lambda t}$ where λ is the rate at which heads appear, or the decay constant of a nucleus. For more, see the *Exponential Distribution* section in the *Probability Distributions* chapter.

Problem 9. If you are betting on a coin toss whose probability of heads varies over time, what is a good strategy to use?

Answer. If the coin's bias is changing, and especially if it has the potential for switching direction, then the BSP (bet same as previous) strategy is good since it can track a switching bias, as long as the switching is not too frequent.

Problem 10. How are the Catalan numbers related to coin tosses?

Answer. If you interpret a series of coin tosses as a random walk on the integers with steps of size $+1$ for heads, and -1 for tails, then the Catalan numbers, $f(n)$, are the number of walks of length $2n$ that return to the starting point for the first time after $2n$ steps and always stay to the right of the starting point.

Problem 11. If your probability of winning a game is

0.6 where you bet $1, what's the minimum bankroll requirement for a 50% chance of getting to $100?

Answer. We use equation 5.6 to get the minimum bankroll amount, $k = (\log(1 + \alpha^N) - \log(2))/\log(\alpha)$ where $\alpha = q/p = 0.4/0.6 = 0.66667$, and $N = 100$ which gives $k = 1.71$. So we need at least about $2 for a 50% chance of getting to $100.

Problem 12. For an unfavorable game playing against an opponent with unlimited resources, what's the average time before you're broke if you win $1 or lose $1 for each round, you start with a bankroll of $10, and your probability of losing ranges from 0.51 to 0.6?

Answer. From equation 5.13, the average duration for an unfavorable game played against an opponent with unlimited resources is $c_k = k/(q - p)$, where $k =$ starting bankroll, $q =$ probability of losing, and $p = 1 - q =$ probability of winning. Here $k = 10, and $q = 0.51, 0.52, \ldots 0.60$, the average time (number of bets) before you're broke is shown in table 9.2.

q	0.51	0.52	0.53	0.54	0.55	0.56	0.57	0.58	0.59	0.6
bets	500	250	167	125	100	83	71	63	56	50

Table 9.2: Average time (number of bets) before you're broke with unfavorable game ($q > 0.5$) against opponent with unlimited resources, starting with bankroll of $10, and betting $1 for each round.

Problem 13. What is the average number of tosses of a fair coin needed to get at least 10 heads in a row?

Answer. This is given in table 6.1, which shows 2,046 tosses for a fair coin ($p = 0.5$) with $r = 10$.

Problem 14. What is the probability that a run of length 5 heads occurs somewhere in the first 25 tosses of a fair coin?

Answer. This can be calculated by use of equations 6.2 and 6.3, which are $f_n = q \sum_{k=1}^{r} p^{k-1} f_{n-k}$ and $F_n = \sum_{i=1}^{n} f_i$ where $p = q = 1/2$, $r = 5$, and we compute the f_n's with initial conditions $f_1 = f_2 = f_3 = f_4 = 0$, $f_5 = 1/32$ to get $F_{25} = 0.3115906$. This result corresponds to the approximate answer indicated in the graph of Figure 6.1, showing a little over 0.3. Note that this calculation is a bit tedious by hand, but is easy with a computer algebra system like Mathematica, Maple, or Maxima.

Problem 15. What is the probability that it takes 50 tosses of a fair coin to get two runs of 5 heads?

Answer. The answer can be gotten by calculating $f_{50}^{(2)}$ using equation 6.34, which is $f_n^{(2)} = f_1 f_{n-1} + f_2 f_{n-2} + \cdots + f_{n-1} f_1$. A more efficient way to get the answer is to square the generating function for f_n, equation 6.17, which is $F(z) = \frac{p^r z^r (1-pz)}{1-z+qp^r z^{r+1}}$ with $r = 5$, $p = q = 1/2$, then take its Taylor series. Our probability is then the coefficient of the z^{50} term of the Taylor series, which we find to be 0.0060427. Note

168

again that a computer algebra system is handy for this problem.

Problem 16. For the previous problem, on average how long does it take to get one run of 5 heads, and two runs of 5 heads?

Answer. The average number of tosses to get one run of 5 heads can be gotten by differentiating, with respect to z, the generating function, $F(z)$, of the previous problem, then evaluating the result at $z = 1$, which gives 62. Similarly, the average number of tosses to get two runs of 5 heads can be gotten by differentiating $F(z)^2$ with respect to z, then evaluating the result at $z = 1$, which gives 124. Note that the average tosses to get two runs is twice that of the average to get one run.

Problem 17. What is the average number of tosses needed to get either 10 heads in a row or 10 tails in a row using a coin with probability of heads $= 0.6$?

Answer. Table 6.6 for $r = 10$ and $p = 0.6$ gives us an answer of 401. This can be calculated by use of equation 6.58, $\mu = \frac{(1-p^h)(1-q^t)}{p^h q + p q^t - p^h q^t}$ with $h = t = 10$, $p = 0.6$, and $q = 0.4$, giving 400.6, as in the table.

Problem 18. What is the probability that a run of 10 heads or 6 tails occurs for the first time on the 50th toss using a coin with probability of heads $= 0.7$?

Answer. We use the generating function of equation 6.57, which is $F(z) = \frac{p^h z^h (1-pz)(1-q^t z^t) + q^t z^t (1-qz)(1-p^h z^h)}{1-z+p^h q z^{h+1} + p q^t z^{t+1} - p^h q^t z^{h+t}}$, evaluate it in a computer algebra system at $h = 10$, $t = 6$, $p = 7/10$ and $q = 3/10$, then take the Taylor series of the result. Our answer is the coefficient of the z^{50} term, which gives 0.00651327.

Problem 19. What is the average number of tosses needed to get either 10 heads in a row or 6 tails in a row using a coin with probability of heads $= 0.7$?

Answer. This can be calculated by use of equation 6.58, $\mu = \frac{(1-p^h)(1-q^t)}{p^h q + p q^t - p^h q^t}$ with $h = 10$, $t = 6$, $p = 0.7$, and $q = 0.3$, giving 108.3275.

Problem 20. What is the probability that it takes 20 tosses of a fair coin to get the pattern HTHTH?

Answer. We are asking for the probability that the pattern occures for the first time after 20 tosses so we can use the Markov matrix in equation 7.7 with $p = q = 1/2$. To get the pattern on the 20th toss we first have to be in state 4 of the automaton in figure 7.5 on the 19th toss and then get one more H. The probability of this is $p \mathbf{M}_{04}^{19} = 0.16696$.

Problem 21. What is the occurence probability of the pattern HTHTH occuring on the 20th toss for a fair coin?

Answer. The occurence probability means that the pattern does not necessarily occure for the first time and that there are no overlaps in the occurences of the pattern. The Markov matrix for this is shown in equation 7.8 with $p = q = 1/2$. To get the pattern we have to start in state 0 and end in state 5 of the automaton in figure 7.6 on the 20th toss. The probability of this is $\mathbf{M}_{05}^{20} = 0.023809$.

Problem 22. For a fair coin, what is the probability that in a sequence of 50 tosses, a head never comes up more than 7 times in a row?

Answer. We start with the run generating function of equation 8.35, $g(a, b) = \frac{(1+a)(1+b)}{1-ab}$ where a represents a run of heads, and b represents a run of tails. To restrict runs of heads to be no more than 7, and no restriction on runs of tails, let $a = z + z^2 + \cdots + z^7$ and $b = z + z^2 + \cdots = z/(1 - z)$. The generating function for the number of tosses with no run of heads greater than 7 is then $g(z + z^2 + \cdots + z^7, z/(1 - z))$. The power series expansion of this function says that there are 1031817231903744 sequences of length 50 with no run of heads greater than 7. The probability is this number divided by 2^{50}, or 0.91644. So it's unlikely we'd get more than 7 heads in a row in 50 tosses of a fair coin.

Problem 23. What is the expected length of the longest run of heads in 50 tosses of a fair coin?

Answer. To get the expectation, evaluate the sum in equation 7.59 up to 50. The coefficient of z^{50} in the power series expansion gives the answer. We get 5.0068.

Problem 24. For a fair coin, what is the probability that the longest run of heads or tails in a sequence of 30 tosses is less than or equal to 5?

Answer. Once again we start with the run generating function of equation 8.35, $g(a,b) = \frac{(1+a)(1+b)}{1-ab}$ where a represents a run of heads, and b represents a run of tails. To restrict runs of heads and tails to be no more than 5, let $a = b = z + z^2 + \cdots + z^5$. The generating function for the number of tosses with no run of heads or tails greater than 5 is then $g(z + z^2 + \cdots + z^5, z + z^2 + \cdots + z^5)$. The power series expansion of this function says that there are 690104702 sequences of length 30 with no run of heads or tails greater than 5. The probability is this number divided by 2^{30}, or 0.64271.

Problem 25. How are the Fibonacci numbers related to coin tosses?

Answer. The Fibonacci numbers come up when counting coin toss sequences where heads never appear two or more times in a row. We can find the generating function for those sequences in the same way as the previous problem but this time with $a = z$ and $b = z/(1-z)$. Substituting this into $g(a,b)$ produces the generating function for the Fibonacci numbers, which is $\frac{1+z}{1-z-z^2}$.

Problem 26. How can you make an unfair coin fair?

Answer. If p is the probability of heads, and $q = 1 - p$ is the probability of tails, then the probability of heads followed by tails is the same as the probability of tails followed by heads. So we use heads followed by tails to denote heads in a fair coin, and tails followed by heads to denote tails in a fair coin. The procedure is then, you flip a coin twice and if you get heads twice or tails twice, then just ignore it, otherwise a heads followed by tails becomes a heads, and a tails followed by heads becomes a tails.

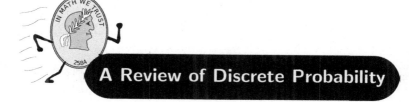

A Review of Discrete Probability

This appendix provides a review of the essentials of discrete probability. It is concerned with observations, experiments or actions that have a finite number of unpredictable outcomes. The set of all possible outcomes is called the sample space (standard terminology) and is denoted by the symbol Ω. An element of Ω (an individual outcome) will be denoted by ω. A coin toss for example, has two possible outcomes: heads (H) or tails (T). The sample space is $\Omega = \{H, T\}$ and $\omega = H$ is one of the possible outcomes. Another example is the roll of a die which has 6 outcomes so that $\Omega = \{1, 2, 3, 4, 5, 6\}$. A subset of the sample space is called an event and is denoted by a capital letter such as A or B. In the die example, let A be the event that an even number is rolled, then $A = \{2, 4, 6\}$.

A.1 Probabilities

Each outcome, ω, will have a probability assigned to it, denoted $P(\omega)$. The probability is a real number ranging from 0 to 1 that signifies the likelihood that an outcome will occur. If $P(\omega) = 0$ then ω will never occur and if $P(\omega) = 1$ then ω will always occur. An intermediate value such as $P(\omega) = 1/2$ means that ω will occur roughly half the time if the experiment is repeated many times. In general, if you perform the experiment a large number of times, N, and the number of times that ω occurs is

$n(\omega)$, then the ratio $n(\omega)/N$ should approximately equal the probability of ω. It is possible to define $P(\omega)$ as the limit of this ratio.

$$P(\omega) = \lim_{N \to \infty} \frac{n(\omega)}{N} \qquad (A.1)$$

The function $P(\omega)$, which assigns probabilities to outcomes, is called a probability distribution. We will now look at some of its defining properties. To begin with, if the probabilities are defined as in equation A.1, then clearly the sum of all the probabilities must equal 1.

$$\sum_{\omega \in \Omega} P(\omega) = 1 \qquad (A.2)$$

It is often necessary to determine the probability that one of a subset of all the possible outcomes will occur. If A is a subset of Ω then $P(A)$ is the probability that one of the outcomes contained in A will occur. Using the definition in A.1 it should be obvious that:

$$P(A) = \sum_{\omega \in A} P(\omega) \qquad (A.3)$$

Many other properties can be derived from the algebra of sets. Let $A + B$ be the set of all elements in either A or B (no duplicates) and let AB be the the set of all elements in both A and B, then:

$$P(A + B) = P(A) + P(B) - P(AB) \qquad (A.4)$$

If A and B have no elements in common then they are exclusive events, i.e. they can not both occur simultaneously. In this case equation A.4 reduces to $P(A + B) = P(A) + P(B)$. In general, the probability that any one of a number of exclusive events will occur is just equal to the sum of their individual probabilities.

A.2 Conditional Probabilities and Independence

Conditional probabilities and the closely related concept of independence are very important and useful in probability calculations. Let $P(A|B)$ be the probability that A has occurred given that we know B has occurred. In short, we will refer to this as the probability of A given B or the probability of A conditioned on B. What $P(A|B)$ really represents is the probability of A using B as the sample space instead of Ω. If A and B have no elements in common then $P(A|B) = 0$. If they have all elements in common so that $A = B$ then obviously $P(A|B) = 1$. In general we have

$$P(A|B) = \frac{P(AB)}{P(B)} \tag{A.5}$$

Using a single fair die roll as an example, let $A = \{1, 3\}$ and $B = \{3, 5\}$ then $AB = \{3\}$, $P(AB) = 1/6$, $P(B) =$

$1/3$ and

$$P(A|B) = \frac{1/6}{1/3} = \frac{1}{2} \qquad \text{(A.6)}$$

Knowledge that B has occurred has increased the probability of A from $P(A) = 1/3$ to $P(A|B) = 1/2$. The result can also be deduced by simple logic. We know that B has occurred therefore the roll was either a 3 or a 5. Half of the B events are caused by a 3 and half by a 5 but only the 3 also counts as an A event also, therefore $P(A|B) = 1/2$.

Conditional probabilities are not necessarily symmetric. $P(B|A)$ need not be equal to $P(A|B)$. Using the definition in equation A.5, you can show that

$$P(A|B)P(B) = P(B|A)P(A) \qquad \text{(A.7)}$$

so the two conditional probabilities are only equal if $P(A) = P(B)$. Another useful thing to keep in mind is that conditional probabilities obey the same properties as nonconditional probabilities. This means for example that if A and B are exclusive events then $P(A + B|C) = P(A|C) + P(B|C)$.

The concept of independence is naturally related to conditional probability. Two events are independent if the occurrence of one has no effect on the probability of the other. In terms of conditional probabilities this means that $P(A|B) = P(A)$. Independence is always symmetric, if A is independent of B then B is independent of

A. Using the definition in equation A.5 you can see that independence also implies that

$$P(AB) = P(A)P(B) \qquad (A.8)$$

This is often taken as the defining relation for independence.

Another important concept in probability is the law of total probability. Let the sample space Ω be partitioned by the sets B_1 and B_2 so that every element in Ω is in one and only one of the two sets and we can write $\Omega = B_1 + B_2$. This means that the occurrence of A coincides with the occurrence of B_1 or B_2 but not both and we can write

$$A = AB_1 + AB_2 = A(B_1 + B_2) = A\Omega \qquad (A.9)$$

The probability of A is then

$$P(A) = P(AB_1) + P(AB_2) \qquad (A.10)$$

This can be extended to any number of sets that partition Ω.

A.3   Random Variables

To carry out any kind of probabilistic analysis we need random variables. A random variable is a bit like the probability distributions discussed above in that it assigns a number to each of the elements in the sample space. It is therefore really more like a function that maps elements in the sample space to real numbers. A random variable is usually denoted with an upper case letter such as X and the values it can assume are given subscripted lower case letters such as x_i for $i = 1, 2, \ldots, n$ where n is the number of possible values. The mapping from an element ω to a value x_i is denoted as $X(\omega) = x_i$. Note that it is not necessary that every element be assigned a unique value and the particular value assigned will depend on what you want to analyze.

A simple example is a coin toss betting game. You guess what the result of the toss will be. If your guess is correct you win \$1 otherwise you loose \$1. The sample space consists of only two elements, a correct guess and an incorrect guess $\Omega = \{\text{correct}, \text{incorrect}\}$. If you are interested in analyzing the amounts won and lost by playing several such games then the obvious choice for the random variable is $X(\text{correct}) = 1$, $X(\text{incorrect}) = -1$. If you are just interested in the number of games won or lost then the random variable $Y(\text{correct}) = 1$, $Y(\text{incorrect}) = 0$ would be better. Often an analysis in terms of one variable can be converted into another variable by finding a relation between them. In the above example $X = 2Y - 1$ could be used to convert between the variables.

As another example consider tossing a coin three times. The sample space consists of 8 elements
$$\Omega = \{TTT, TTH, THT, THH, HTT, HTH, HHT, HHH\}$$
where T indicates the toss was a tail and H a head. This time we let X be the random variable that counts the

number of heads in the three tosses. It can have values 0, 1, 2, or 3 and not every element in the sample space has a unique value. The values are $X(TTT) = 0$, $X(TTH) = X(THT) = X(HTT) = 1$, $X(THH) = X(HTH) = X(HHT) = 2$, $X(HHH) = 3$.

Probability distributions are most often expressed in terms of the values that a random variable can take. The usual notation is

$$P(X = x_i) = p(x_i) \qquad \text{(A.11)}$$

The function $p(x_i)$ is the probability distribution for the random variable X. It is often also called the probability mass function. Note that it is not necessarily the same as the probability distribution for the individual elements of the sample space since multiple elements may be mapped to the same value by the random variable. In the three coin toss example, each element in the sample space has a probability of $1/8$, assuming a fair coin. The probability distribution for X however is $p(0) = 1/8$, $p(1) = 3/8$, $p(2) = 3/8$, $p(3) = 1/8$. It will always be true that the sum over all the probabilities must equal 1.

$$\sum_i p(x_i) = 1 \qquad \text{(A.12)}$$

A.4 Expectation and Variance

The two most important properties of a random variable are its expectation (also called mean) and variance. The expectation is simply the average value of the random variable. In the coin toss betting game, X can have a value of $+1$ or -1 corresponding to winning or losing. In N flips of the coin let k be the number of wins and $N - k$ the number of losses. The total amount won is then

$$W = k - (N - k) \tag{A.13}$$

and the average amount won per flip is

$$\frac{W}{N} = \frac{k}{N} - (1 - \frac{k}{N}) \tag{A.14}$$

As the number of flips becomes very large the ratio k/N will equal $p(1)$, the probability of winning, and the equation then becomes equal to expectation of the random variable.

$$E[X] = p(1) - p(-1) \tag{A.15}$$

Where $p(-1) = 1 - p(1)$ is the probability of losing and $E[X]$ is the usual notation for the expectation of X. In this case the expectation is the average amount that you can expect to win per flip if you play the game for a very long time.

In general if X can take on n values, x_i, $i = 1, 2, \ldots, n$ with corresponding probabilities $p(x_i)$ then the expectation is

$$E[X] = \sum_{i=1}^{n} p(x_i)x_i \qquad \text{(A.16)}$$

The expectation gives you the average but in reality large deviations from the average may be possible. The variance of a random variable gives a sense for how large those deviations can be. It measures the average of the squares of the deviations. The equation for the variance is:

$$\text{Var}[X] = \sum_{i=1}^{n} p(x_i)(x_i - E[X])^2 \qquad \text{(A.17)}$$

The equation simplifies somewhat to

$$\text{Var}[X] = E[X^2] - E[X]^2 \qquad \text{(A.18)}$$

where

$$E[X^2] = \sum_{i=1}^{n} p(x_i)x_i^2 \qquad \text{(A.19)}$$

is the expectation for the square of the random variable.

In general the expectation for any function $g(X)$ is:

$$E[g(X)] = \sum_{i=1}^{n} p(x_i)g(x_i) \qquad \text{(A.20)}$$

Another useful measure of deviation from the average is called the standard deviation, σ. It is found by taking the square root of the variance.

$$\sigma = \sqrt{\mathrm{Var}[X]} \qquad \text{(A.21)}$$

A.5   Probability Generating Function

We can always map the values of a discrete random variable to the integers $k = 0, 1, 2, \dots$. With these values we can define a probability generating function $G(z)$ such that the coefficient of z^k in its power series expansion is the probability that $X = k$. In other words, if p_k is the probability that $X = k$, then the power series expansion of $G(z)$ looks like

$$G(z) = \sum_{k=0}^{\infty} p_k z^k \qquad \text{(A.22)}$$

One of the biggest advantages of the generating function is that it can be used to calculate the expectation and the variance. The first derivative evaluated at $z = 1$ produces the expectation, $E[X]$.

$$E[X] = \frac{dG}{dz}\bigg|_{z=1} = \sum_{k=1}^{\infty} p_k k \qquad \text{(A.23)}$$

To get the variance we need $E[X^2]$ which is given by:

$$E[X^2] = \frac{d}{dz}\left(z\frac{dG}{dz}\right)\bigg|_{z=1} = \sum_{k=1}^{\infty} p_k k^2 \qquad \text{(A.24)}$$

Using these two expressions in equation A.18 produces the variance, $\text{Var}[X]$.

A.6   Joint Probability Distribution

As we saw above, a sample space can have more than one random variable defined on it. If we have two variables X and Y then we can define the probability that $X = x_i$ at the same time that $Y = y_j$. This is called the joint probability distribution for X and Y.

$$P(X = x_i, Y = y_j) = p(x_i, y_j) \qquad \text{(A.25)}$$

The individual distributions, $p(x_i)$ and $p(y_j)$, are recovered by summing the joint distribution over one of the variables. To get $p(x_i)$ you sum $p(x_i, y_j)$ over all the possible values of Y.

$$p(x_i) = \sum_{j} p(x_i, y_j) \qquad \text{(A.26)}$$

and likewise for $p(y_j)$

$$p(y_j) = \sum_i p(x_i, y_j) \qquad \text{(A.27)}$$

From these last two equations it is obvious that if you sum over both variables of the distribution, the result should equal 1.

$$\sum_i \sum_j p(x_i, y_j) = 1 \qquad \text{(A.28)}$$

It is possible to construct a joint distribution for any number of random variables, not just 2. For example $p(x_i, y_j, z_k)$ would be a joint distribution for the variables X, Y, and Z.

With a joint distribution you can calculate the expectation and variance for functions of variables. The expectation for the sum $X + Y$ is:

$$
\begin{aligned}
E[X + Y] &= \sum_i \sum_j p(x_i, y_j)(x_i + y_j) \qquad \text{(A.29)} \\
&= \sum_i x_i \sum_j p(x_i, y_j) + \sum_j y_j \sum_i p(x_i, y_j) \\
&= \sum_i x_i p(x_i) + \sum_j y_j p(y_j) \\
&= E[X] + E[Y]
\end{aligned}
$$

The property that the expectation for a sum of variables

is equal to the sum of their expectations is called linearity and it is true for the sum of any number of variables. For three variables for example $E[X + Y + Z] = E[X] + E[Y] + E[Z]$. Another easily verifiable consequence of linearity is that for any constants a and b

$$E[aX + bY] = aE[X] + bE[Y] \qquad \text{(A.30)}$$

In the example of the coin toss game we had two random variables that were related by $X = 2Y - 1$. The linearity property of the expectation means that $E[X] = 2E[Y] - 1$, where we used the fact that the expectation of a constant is just the constant.

The expectation for the product XY is

$$E[XY] = \sum_i \sum_j p(x_i, y_j) x_i y_j \qquad \text{(A.31)}$$

If the variables X and Y are independent then the joint distribution can be factored into a product of the individual distributions, $p(x_i, y_j) = p(x_i)p(y_j)$. In this case you can show that the expectation of the product is the product of the expectations, $E[XY] = E[X]E[Y]$.

For the variance of a sum we have

$$\text{Var}[X + Y] = E[(X - E[X] + Y - E[Y])^2] \qquad \text{(A.32)}$$

after expanding and simplifying this becomes

$$\mathrm{Var}[X + Y] = \mathrm{Var}[X] + \mathrm{Var}[Y] + 2\mathrm{Cov}[X, Y] \quad (\text{A.33})$$

where $\mathrm{Cov}[X, Y]$ is called the covariance of X and Y. The covariance is defined as:

$$\mathrm{Cov}[X, Y] = E[XY] - E[X]E[Y] \quad\quad (\text{A.34})$$

For independent variables the covariance is zero. The variance of the sum is then just the sum of the variances.

Further Reading

- *Probability Problems and Solutions*, Hollos and Hollos
  Helps you learn probability through problem solving.

- *Combinatorics Problems and Solutions*, Hollos and Hollos
  Helps you learn combinatorics through problem solving.

- *Combinatorics II Problems and Solutions: Counting Patterns*, Hollos and Hollos
  A continuation of our first combinatorics book. Teaches powerful methods for counting patterns.

- *Finite Automata and Regular Expressions: Problems and Solutions*, Hollos and Hollos
  Helps you learn automata and regular expressions through problem solving.

- *Probability Tales*, Grinstead, Peterson and Snell
  Contains a lot of material related to coin tossing.

- *An Introduction to Probability Theory and its Applications*, William Feller
  A classic book on probability that covers many results related to coin tossing.

- *Heads or Tails: An Introduction to Limit Theorems in Probability*, E. Lesigne
  Discusses limit theorems (concerned with repeated experiments with finite outcomes) from the perspective of the coin toss.

- *Runs and Scans with Applications*, Balakrishnan and Koutras
  Covers runs in binary sequences.

- *Distribution Theory of Runs and Patterns*, Fu and Lou
  Covers the Markov chain approach to runs and patterns in binary sequences.

- *The Longest Run of Heads*, Mark F. Schilling
  The College Mathematics Journal, Vol. 21, No. 3, (1990), pp. 196-207.
  Summary: Development of recursion formulas that generate the exact distribution of the largest run of heads and several curious features of head run distributions.

Stefan Hollos and **J. Richard Hollos** are physicists and electrical engineers by training, and enjoy anything related to math, physics, engineering and computing. In addition, they enjoy creating music and visual art, and being in the great outdoors. They are the authors of:

- **Creating Melodies**

- **Hexagonal Tilings and Patterns**

- **Combinatorics II Problems and Solutions: Counting Patterns**

- **Information Theory: A Concise Introduction**

- **Recursive Digital Filters: A Concise Guide**

- **Art of Pi**

- **Creating Noise**

- **Art of the Golden Ratio**

- **Creating Rhythms**

- **Pattern Generation for Computational Art**

- **Finite Automata and Regular Expressions: Problems and Solutions**

- **Probability Problems and Solutions**

- **Combinatorics Problems and Solutions**

- **The Coin Toss: Probabilities and Patterns**

- **Pairs Trading: A Bayesian Example**

- **Simple Trading Strategies That Work**

- **Bet Smart: The Kelly System for Gambling and Investing**

- **Signals from the Subatomic World: How to Build a Proton Precession Magnetometer**

They are brothers and business partners at Exstrom Laboratories LLC in Longmont, Colorado. Their website is exstrom.com

Acknowledgements

We'd like to thank our parents for everything that we have.

We thank the makers and maintainers of all the software we've used in the production of this book: the Emacs text editor, the Latex typsetting system, Imagemagick, Inkscape, POV-Ray, Gnuplot, Evince, Maxima, gcc, Bash shell, and the Linux operating system.

Lewis & Clark Bicentennial Silver Dollar Coin.
Image credit: United States Mint.

Thank You

Thank you for buying this book.

If you'd like to receive news about this book and others published by Abrazol Publishing, just go to

http://www.abrazol.com/

and sign up for our newsletter.